Bioethics Yearbook

VOLUME 1
THEOLOGICAL DEVELOPMENTS IN BIOETHICS:
1988-1990

Bioethics Yearbook

VOLUME 1

Bioethics Yearbook

VOLUME 1
THEOLOGICAL DEVELOPMENTS IN BIOETHICS:
1988-1990

THE CENTER FOR ETHICS, MEDICINE AND PUBLIC ISSUES
Baylor College of Medicine
The Institute of Religion
Rice University
Houston, Texas, U.S.A.

Edited by

Baruch A. Brody, *Director*
B. Andrew Lustig, *Senior Editor for this Volume*
H. Tristram Engelhardt, Jr.
Laurence B. McCullough

Thomas J. Bole, III, *Managing Editor*

KLUWER ACADEMIC PUBLISHERS
DORDRECHT / BOSTON / LONDON

ISBN 0-7923-1280-5

ISSN 0926-261X

Published by Kluwer Academic Publishers,
P.O. Box 17, 3300 AA Dordrecht, The Netherlands.

Kluwer Academic Publishers incorporates
the publishing programmes of
D. Reidel, Martinus Nijhoff, Dr W. Junk and MTP Press.

Sold and distributed in the U.S.A. and Canada
by Kluwer Academic Publishers,
101 Philip Drive, Norwell, MA 02061, U.S.A.

In all other countries, sold and distributed
by Kluwer Academic Publishers Group,
P.O. Box 322, 3300 AH Dordrecht, The Netherlands.

Printed on acid-free paper

Printed in the Netherlands

TABLE OF CONTENTS

INTRODUCTION

It is with pleasure and pride that the Center for Ethics, Medicine, and Public Issues offers this initial volume in our series of Bioethics Yearbooks. In conjunction with Kluwer Publishers, we plan a yearbook series which will alternate between a biannual volume tracing recent theological discussion in bioethics and a biannual volume tracing recent governmental bioethics policies. Each yearbook will draw upon the acknowledged expertise of respected commentators, who will chronicle, as well as interpret, recent concerns in bioethics in their specific areas of competence.

In this inaugural volume, we have invited twelve scholars to summarize bioethical discussions in twelve religious traditions from 1988-1990 on the following list of topics: new reproductive technologies, abortion, maternal-fetal conflicts, care of severely disabled newborns, consent to treatment and experimentation, confidentiality, equitable access to health care, ethical concerns raised by cost-containment measures, decisions to withhold or withdraw life-sustaining treatment, active euthanasia, the definition of death, and organ donation and transplantation. Each author addressed only those issues which have been recently discussed in his or her tradition. Authors were also free to comment upon other issues not on the list.

The present yearbook is an effort to meet the need for a single volume that provides comprehensive and timely summaries of discussions in a wide variety of religious traditions on specific topics in bioethics. We have commissioned expert commentators to trace recent trends and developments in their respective traditions – both formal statements by official bodies and study groups, and less formal discussions that reflect currents of thought in the ongoing dialogue on these issues. Especially for this first volume in the yearbook series, we have left it to the discretion of our authors to place recent developments in a broader context by including those background documents and discussions they deem necessary to clarify recent concerns.

In addition to providing up-to-date summaries of recent developments, our authors have also offered rich and useful bibliographical references to a wide array of documents, many of which would otherwise prove difficult for individuals to know about, much less to assemble. We invite our readers to contact us with their advice about ways to improve and expand the range of theological resources upon which our authors may draw in future yearbooks.

As is always the case with a volume of this kind, the overall quality of the yearbook depends primarily on the quality of the individual efforts of its contributors. The scholarship of our authors will be evident. But in less obvious ways, the professional courtesy and responsibility of our authors, evidenced by their unanimous cooperation in meeting all deadlines, made the editorial oversight of this volume uncharacteristically easy. Finally, we owe special thanks to Thomas J. Bole, III, for his work as managing editor, to Sarah Vaughan

Brakman, for her welcome help with the final editing chores, and to Delores D. Smith, for her exacting attention to the countless details that go into producing a complex volume such as this.

B. Andrew Lustig, Senior Editor
Baruch A. Brody, Editor, Director of the Center
H. Tristram Engelhardt, Jr., Editor
Laurence B. McCullough, Editor

FOREWORD TO VOLUME I

No one needs to argue for the legitimacy and value of a book on bioethics which incorporates diverse religious perspectives. The very field of study called bioethics grew out of religiously motivated concerns less than thirty years ago. Prior to the 1960s, Protestant theologians had given hardly a serious thought to questions of medical ethics, although the Roman Catholic writer Gerald Kelly had already systemized the teaching derived from a tradition in his communion. The Jewish and Islamic traditions of morality in cases of childbirth, sickness, and dying likewise had a long history.

The sudden advent of technological devices and sophisticated methods in both biological research and medical practice alerted certain theologians to the opening of unprecedented problems for moral counsel and decision-making. Transplanting human organs, prenatal testing of the unborn, securing informed consent of subjects of experimentation, and opening the secrets of the DNA molecule were among the rapid developments which stimulated interest. The Methodist Paul Ramsey and the Episcopalian Joseph Fletcher, though driven by very different theological perceptions and convictions, became dominant, much-quoted leaders in the new discipline. Daniel Callahan, at that time a noted Catholic lay theologian, and Richard A. McCormick, S. J., widened the ecumenical dimensions of the field. Some of the best writing was done by them and found in religious journals.

As more and more philosophical, legal, and sociological implications of modern medicine and biotechnology became evident, the influence of writers who claimed no religious faith became ascendant. Some persons of specific religious identity who entered the field were apparently so impressed by the rise of secular rules of logic and ethical discrimination that they virtually concealed their beliefs in divinely willed and traditionally taught criteria of judgment.

Religious thinkers do not function individualistically. They belong to certain communities of faith and are inevitably participants in communal efforts to articulate and make public their ethical positions on specific problems. Various religious communities have thus been shaping and refining their particular teachings on bioethical issues. Some of these are expressed in canonical instructions to their constituents; others are hortatory, or intended only to serve as favored guidelines for individuals.

Since bioethics has become a sphere of interest to large numbers of people, it is urgently important to make these religious teachings accessible to a widening public. Documentation is often hard to find. Statements are buried in the minutes and resolutions of church assemblies and synods, or available only in small, fugitive brochures. Some gain so little currency that they remain almost clandestine. Yet there are many people who are eager to read them. The present book serves not only to provide information about the various religious traditions, but also offers helpful, elucidating interpretations by responsible representatives. Documents without interpretation are less than fully useful.

The diversities of viewpoint, belief, and judgment found among and within the religious communities will possibly be a cause for some readers' disappointment. This may occur if they hold a conventional view about religious teachings on ethical issues: namely, the view that every church, religious community, or denominational body has a set of specific, authoritative principles and rules. Journalists who want to satisfy conventional minds, or who have such themselves, frequently ask, "What is *the* Christian teaching on organ transplants?", or, "What is *the* position of Muslims or Methodists on gene therapy, fetal tissue experimentation, or patenting modified animals?". They want easy answers to hard, sometimes insoluble, problems. While celebrating freedom of thought and conscience, they expect religious groups to be monochromatic and dogmatic. Some are; most are not.

The conventional mind also errs in assuming that each religious body teaches with the same sense of authority over its adherents. In fact, the profile of authoritative policies is very irregular, reaching from the *magisterium* of the Roman Catholic Church to the good counsel given by some Protestant assemblies.

Not only these irregularities, but the wide diversity of the teachings as such constitute an emerging problem in the ecumenical movement of Christians, as well as for more inclusive interfaith relationships. While the rule of "diversity within unity" is honored, especially where ultimate issues of human life are concerned, there are some diversities which are really divisive. Bioethics for some has become a more formidable obstacle to religious unity than traditional doctrines. How to weigh the social value of unity against the quest for what is true constitutes a vexed question today. To help people in this deliberation, it is necessary to have authentic, dependable, up-to-date information and interpretation of the movements of thought within the communities represented. Herein lies the practical value of this collection of essays for teachers, physicians, nurses, clergy, lawyers and makers of public policy.

The Center for Ethics, Medicine, and Public Issues was constituted in 1982 by three sponsoring institutions in Houston: Rice University, Baylor College of Medicine, and the Institute of Religion in the Texas Medical Center. Among the four distinguished members of the Center's faculty, B. Andrew Lustig, Ph.D., has the specific task of studying and interpreting theological ethics concerning biomedical matters. Because he is also a faculty member of the Institute of Religion, seconded to the Center, it gives me pleasure to commend him for his outstanding effort as senior editor of this excellent volume.

J. Robert Nelson, D.Theol.
Director
The Institute of Religion

JOSEPH BOYLE

THE ROMAN CATHOLIC TRADITION AND BIOETHICS

I. INTRODUCTION

Roman Catholic writing on bioethics comprises an extensive and sprawling body of literature authored not only by established authorities such as popes, bishops, and their advisors, but also by a considerable body of theologians and philosophers who may or may not have official positions within the Church. What distinguishes this literature from other bioethical writing is its explicit dependence on the Roman Catholic moral tradition. This dependence usually involves acceptance of large parts of received Roman Catholic teaching and use of the categories of moral analysis developed within the tradition. But it frequently involves criticism, refinement, and development of traditional teaching on some points.

The published work, even during a two-year period, of a large community with an established and robust tradition of ethical commentary is too extensive to summarize in a short report, and too extensive for one reporter to assimilate. So, considerable selection has been necessary.

I have been selective in these ways: First, I have focused almost entirely on materials readily available in English. English-speaking bishops and moralists have contributed, both quantitatively and qualitatively, to bioethical discussion disproportionately to their numbers within the Church. Moreover, virtually all ecclesiastical documents are available in authorized English translation, and the literature in English is sensitive to, and continuous with, the discussion in French, German, Italian, and Spanish.

Second, I have focused on areas in which there have been significant doctrinal developments of specifically bioethical interest. Thus, I have avoided areas in which received teachings are simply repeated, or in which the developments are not specifically bioethical, or in which Catholic statements simply reflect more generally held views.

In particular, this principle for selection excludes much of the Catholic writing on abortion, suicide, and euthanasia, which takes for granted the received moral teaching and deals with political and disciplinary implications which are of secondary interest for bioethics. It also excludes the considerable discussion of the authority for Catholics of the various Church teachings relevant to bioethics, and the important debates now going on among Catholics about ethical theory and moral methodology.

Finally, I note several areas of overlap between the organization of this report and the division of topics used throughout this volume. First, although I do not have a section on abortion, I report what I think is the most important

Catholic development related to abortion, the discussion of the status of the embryo during the first weeks after conception, in the section on reproductive technologies. Second, I report the discussion of the transplantation of tissue from fetal remains in the section on consent to experimentation, because the issue in this discussion is not specifically about the ethics of transplantation.

II. NEW REPRODUCTIVE TECHNOLOGIES

The "Instruction on Respect for Human Life in its Origins and on the Dignity of Procreation" is by far the most important document on bioethical matters issued by the Roman Catholic Church in recent years. This document, issued by the Congregation for the Doctrine of the Faith and approved by Pope John Paul II on February 22, 1987, is usually called *Donum vitae*, after the first two words of the official Latin text. *Donum vitae* deals with a number of questions concerning respect for embryos, and with many of the moral questions raised by techniques of artificial reproduction.[1] Furthermore, it is not simply a disciplinary decree but moral teaching, and so presents some of the reasoning behind its traditional conclusions.

Donum vitae received considerable critical attention when it appeared, not only in Catholic circles, but also in the popular media ([20]; [22]; [33], pp. 118-126; [39], pp. 111-131). A number of the more reflective reactions by Catholic theologians appeared during the period covered in this yearbook, and comprise much of the Catholic theological discussion of reproductive technology going on then.

Donum vitae's best known and most controversial evaluations are those dealing with *in vitro* fertilization (IVF) and similar techniques. The theological discussion of these evaluations will be the focus of this summary.

Donum vitae judges IVF to be immoral, both because of the various threats it poses to embryonic life, and because it is judged, as artificial insemination has been judged by the Roman teaching authorities for almost a century, to be an improper way to procreate.

The Status of the Early Embryo

The first of these lines of reasoning is based on the conviction that the human embryo should be respected as a person. *Donum vitae* concludes from this conviction that IVF, as usually practiced, is morally unacceptable because it involves the planned destruction of some embryos, and because it puts embryos at unjustified risk ([10], p. 703). Although concerns exist about the exact nature and extent of the risks it is reasonable to impose upon embryos ([39], pp. 121-122), no one has challenged *Donum vitae*'s basic inference from this conviction to its negative evaluation of the standard practice of IVF. But some Catholic moralists have cautiously questioned the conviction itself ([26], pp. 343-344).

Donum vitae's conviction that even very early embryos should be treated as persons is founded on the presumption, which it holds to be indefeasible, that personal existence begins at conception. This is a presumption because it is not possible to establish when the human soul is present; it is indefeasible because no further scientific or philosophical evidence could establish that the human individual which begins at conception is not personal. The argument is elliptical and carried by a rhetorical question: "How could a human individual not be a human person?" ([10], p. 701).[2]

This presumption plainly presupposes that there is a new human individual from conception. And it is this presupposition that some have questioned. The most developed analysis of this issue is that of Norman Ford, an Australian philosopher who fully accepts the Church's moral condemnation of abortion ([12], p. 62), but who believes that the Church's pro-life stance is undermined by its reliance on an assumption which lacks biological and philosophical support ([12], p. 64).

Ford argues that an ontologically human individual comes to be at the primitive streak stage, about fourteen days after conception. He admits that before the primitive streak stage the human zygote is biologically human, and in some sense of the term is an "individual", but argues that the individual life of any developed human being begins at this point and cannot reasonably be said to have begun before it.

Ford's central argument is too complex to summarize fairly in a brief way. It depends essentially on considering the relevant biological facts in the light of an analysis of the notion of individuality. It concludes that only on pain of accepting a host of absurdities and implausibilities can we regard as the same individual human being the zygote prior to the primitive streak stage and any more developed human who comes to be from it. Ford makes use of the familiar examples of twinning and of the possibility of distinct embryonic cells combining to form a chimeric individual before the primitive streak stage to generate some of the absurdities ([12], pp. 132-163). But his central, and novel, argument appears to be that only at the primitive streak stage does the zygote acquire features we regard as necessary for being an individual human organism, for example, having a definite shape with recognizable boundaries and identifiable parts ([12], pp. 162, 170-177).

Ford's analysis is not without difficulties ([15], pp. 35-40; [37], pp. 106-111; [38]), but the issues he raises certainly deserve further consideration. In particular, Ford's analysis and the criticisms of it suggest that there may be more than a single coherent notion of what it is to be a human individual. Unless these notions are clearly distinguished and their possibly quite different moral implications explored, the established Catholic presumption that personal existence begins at conception is likely to appear an unfounded assumption.

IVF and the Dignity of Procreation

Donum vitae's second line of reasoning – that IVF and similar techniques are improper ways to procreate – has received much more critical comment than the first. Indeed, the Catholic discussion of reproductive technologies has become highly focused on one application of this line of reasoning: the use of IVF and artificial insemination *within* marriage, that is, on cases where the spouses provide the gametes and the genetic mother bears the child. For even those Catholic moralists who take strong exception to some aspects of *Donum vitae* generally accept the idea that human reproduction is something which should take place within marriage, so that the use of third party gametes is morally unacceptable ([26], pp. 340-346; [33], p. 138; cf. [20], 213-216).

The center of the discussion is *Donum vitae*'s claim that procreation and marital lovemaking should not be separated. This dictum, often called the "inseparability principle", is well established in twentieth-century Church teaching concerning both the morality of contraception and of artificial reproduction ([19], pp. 14-17; [26], pp. 333-340): contraception separates lovemaking from procreation by excluding the possibility of pregnancy, and techniques of artificial reproduction such as artificial insemination and IVF separate them by seeking a child but not as the result of an act of marital intercourse. Thus, the plain sense of the principle, as relevant to techniques of artificial reproduction, is that children should come to be *only* through acts of marital intercourse. Hence, all methods of artificial reproduction which replace sexual intercourse as the cause of human generation are judged immoral ([10], pp. 706-707).

The critical question, therefore, is why precisely this kind of separation of marital lovemaking and reproduction is wrongful. Why is an act of marital intercourse the only morally acceptable way to bring human beings into existence? Catholic critics of *Donum vitae* argue that there is no good answer to this question, while its defenders seek to show that there is.

These critics acknowledge that procreation should be within marriage and should be essentially connected with the loving union and shared life of the spouses; they also agree that artificial reproduction is easily abused, and is generally less desirable than natural procreation ([7], pp. 119-152; [26], pp. 347-349; [33], pp. 107-126). But they argue that the inseparability principle overstates and misapplies these common Catholic value judgments.

Shannon and Cahill argue that *Donum vitae* misapplies these value judgments because it places too great an emphasis on discrete actions instead of contextualizing reproduction within the entire married life of a couple. It is not, they maintain, within isolated acts of reproduction or lovemaking that these underlying values are realized or harmed, but within the context of a couple's whole common life. Shannon and Cahill recognize that this way of understanding the relationship among actions, values, and people's lives is contrary to the natural law approach assumed by *Donum vitae*; but they argue that a defensible approach to natural law must pay far more attention to the experience of those involved in the issues being evaluated than the traditional approach allows. Furthermore, the overwhelming experience of Catholic married couples confirms not *Donum vitae*'s approach to the issues of procreation, but the more holistic approach Shannon and Cahill favor ([33], pp. 111-115, 127-139; cf. [7], pp. 144-146; [20], p. 218).

This argument plainly raises large questions of ethical theory, which, if answered in the way Shannon and Cahill think they should be, would render obsolete the reasoning and conclusions of *Donum vitae*. Still, it seems most appropriate in the context of this report to take *Donum vitae* on its own terms, and to consider whether, given its (admittedly controversial) assumptions about ethical theory and moral reasoning, its conclusions follow from its premises.

Shannon's and Cahill's line of argument does not do this. For it appears to slip away from the central issue raised by the inseparability principle: Is there anything about the dissociation of reproduction from a given act of marital intercourse which morally flaws any act of reproduction which involves that dissociation? An argument, like Shannon's and Cahill's, which denies that such a flaw should be morally decisive, does not help us to see whether or not such a flaw is present.

Richard McCormick's discussion appears to have the question posed in the previous paragraph clearly in mind. Two of his observations constitute an argument for a negative answer. First, he reports that he has discussed the question of artificial reproduction within marriage with a variety of thoughtful people, and that none of them accepts the position of *Donum vitae*; the context makes clear that the inseparability principle is precisely what these people do not accept. Second, he notes that the germ of truth in *Donum vitae*'s inseparability principle is

> ...arguably a legitimate aesthetic or ecological (bodily integrity) concern. All artificial interventions, whether to promote or to prevent conception, are a kind of "second best." They involve certain disvalues that, absent sterility or fertility, respectively, we would not entertain. In this sense we can agree with the Congregation that conception achieved through in vitro fertilization is "deprived of its proper perfection." Much the same can be said of contraception. However, a procedure "deprived of its proper perfection" is not necessarily wrong in all cases – unless we elevate an aesthetic-ecological concern into an absolute moral imperative ([26], p. 348; cf. [19], p. 14).

In effect, McCormick challenges *Donum vitae* and its defenders: "You say that the dissociation of procreation from marital intercourse morally flaws any reproductive act which separates them. But I have looked for the moral flaw and do not find it. What I find instead are undesirable features of such acts, which sometimes perhaps, but not inevitably, flaw them morally. Please exhibit the moral flaw in separation as such." This is a powerful challenge.

Donum vitae shows some recognition of this challenge. For it offers reasons why the inseparability principle should be applied in the case of reproductive technology, and so does not suppose that it is an ultimate or obvious principle. Similarly, contrary to McCormick's suggestion ([26], p. 347), *Donum vitae* does not regard the application of this principle to reproductive technologies as a simple logical implication of its use by recent popes in their arguments against contraception.

The general idea behind the inseparability principle seems to be that the values of marital lovemaking and procreation are so intertwined and interdependent that acts which separate them will harm both values ([20], pp. 220-228; [17], pp. 409-410; [21], pp. 69-70). The way these values are harmed in contraception is likely to be different than the way they are harmed in reproductive techniques which separate them, and *Donum vitae* seems to recognize this by indicating, although only very briefly, how these techniques harm these values. *Donum vitae*'s chief concern in this regard seems to be that reproduction which occurs outside marital intercourse fails to respect the equal dignity of the person who is coming to be. One who comes to be, not as the fruit of his parents' love, but as the product of an intervention of medical or biological techniques, would be reduced to the status of an object of scientific technology, and that presumably is inconsistent with equal human dignity ([10], p. 706).

So, *Donum vitae* points to a moral flaw in acts of reproduction which are not acts of marital intercourse (a flaw which, it should be noted, might also be present in some reproductive acts done through marital intercourse). But this indication is hardly an adequately developed moral argument, as some defenders of *Donum vitae* implicitly acknowledge ([23], pp. 34-36; [24], pp. 83-85; [17], pp. 410-411; [21], pp. 70-72). I am not aware that there is a fully and carefully developed argument along these lines.

The development of such an argument would be difficult. The controversy suggests some of the conditions which such an argument would have to meet. First, such an argument would require a definition of the notion of manufacture or production as applied to a biological area, in which living materials, although manipulated and brought together in various ways, remain irreducible causal factors in the outcomes ([26], p. 337). Secondly, such an argument would need to show that the outcome of this kind of technological process is correctly understood as a product – that is, as something necessarily subordinate to those controlling the process. Thirdly, such an argument would have to show that the subordination involved is morally objectionable. To do this, it would be necessary to show that human procreation can proceed in some different way – that at least some acts of procreation by way of marital intercourse can treat the person coming to be differently than reproductive technologies do. In other words, more work is needed to show that procreation by marital intercourse and artificial reproduction have contrary, not complementary, moral features ([19], p. 15; [39], pp. 115-116). Why are children who come to be as the fruit and blessing of their parents' union in the single flesh of marital intercourse not products in just the same way as those who come to be as the result of technological intervention? Or why are those who come to be through reproductive technology not also gifts and blessings which supervene upon their parents' love?

In short, the discussion during the past few years of *Donum vitae*'s argument concerning the dignity of procreation is not complete. But it has led to a more precise formulation of the issues in controversy, sufficient to allow a more focused, and perhaps, definitive discussion to begin.

III. CONSENT TO TREATMENT AND EXPERIMENTATION

Donum vitae touched briefly upon the morality of experimentation upon human embryos and fetuses. Embryos and fetuses are regarded as persons, and so must be respected in experimental contexts just like other people. So, whether viable or not, and whether in utero or not, they have rights, and so may not be treated like subhuman life forms. Experimental therapies, therefore, can be used for their benefit in circumstances in which such therapies would be appropriate for other people. But nontherapeutic experiments (which *Donum vitae* distinguishes very roughly from experiments which are non-invasive) are ruled out. For the risks of experimentation to the embryo or fetus are always significant, and the parents do not have the authority to give informed consent: "The informed consent ordinarily required for clinical experimentation on adults cannot be granted by the parents, who may not freely dispose of the physical integrity or life of the unborn child" ([10], p. 703).

Donum vitae takes the same general approach to the treatment of the remains of embryos and fetuses. They should be respected just like other human remains. It states five necessary conditions for the morally justified use of the corpses of unborn humans: (1) death must be verified; (2) informed consent of the parents must be obtained; (3) complicity in deliberate abortion must be avoided; (4) the risk of scandal must be avoided: and (5) there must be no commercial trafficking in fetal remains ([10], p. 703).

Given the practically important questions these teachings raise, it is surprising that there has been so little discussion of them by Catholic moralists. Daniel Sulmasy is one of the few to have addressed these questions (cf. also [3]); he does so with specific reference to the use of fetal brain tissue for transplant to people suffering from Parkinson's disease. He notes that persons working within the Catholic moral tradition can favor the use of fetal tissue for transplant, but only if certain limits are respected. One of these limits is set by the traditional view that the power of people to dispose of their lives and bodily functions is limited. Another is set by what he calls the "principle of equality", which requires equal treatment of fetuses going to term and fetuses to be aborted. He appears to take this principle to exclude extracting vital brain tissue from any living fetus ([36], pp. 101-102).

When considering the possibility of using tissue from dead fetuses, Sulmasy's position is more permissive. He appears to think that the two major obstacles to doing this when the fetus has been deliberately aborted can be overcome. These obstacles are: the belief that the status of the dead fetus as a victim of a crime overturns the standard presumption that cadavers may be used for such purposes, and the genuine difficulty of obtaining appropriate informed consent.

With respect to the first, he argues that even cautious moralists, following *Donum vitae*'s prescriptions, could accept his position. First, he notes that these moralists regard dead fetuses as human corpses, and accept as established the moral legitimacy of organ donation from cadaveric sources. Thus they must allow such use of fetal tissue as permissible. And the fact that a fetus was killed by illegitimate abortion does not alter this conclusion. For it is permissible to - use the corpse of a murdered adult as a source of donated organs ([36], p. 102). Later he says that, if appropriate informed consent can be obtained, accepting a transplant "would be no more an act of approval of abortion than accepting a liver transplant from the body of a murdered man is an act of approval of murder" ([30], p. 104; cf. [3], p. 319).

Sulmasy is surely correct in saying that such use of the cadavers of aborted fetuses does not imply approval of deliberate abortion. One who makes use of the tissue of a wrongfully aborted fetus need not want or will that the fetus be aborted. But this is only one, very strong, form of complicity in abortion. Its absence does not guarantee that there is no other morally objectionable complicity.

James Burtchaell, in an explanation of the University of Notre Dame's policy excluding the experimental use of tissue from the remains of deliberately aborted fetuses, usefully distinguishes four kinds of moral complicity: "direct and active participation, association that fails to disentangle itself, dereliction of the duty to supervise, and protective assistance" ([8], p. 9). Burtchaell maintains that the scientist who makes use of the tissue of aborted fetuses is involved in the second sort of complicity with the abortionist. He maintains that this sort of complicity makes the researcher a party after the fact to the destruction of the unborn ([8], p. 8). This may not contribute to the causing of the abortion, but it does corrupt the researcher. For the researcher does benefit directly from the abortionist's action: "A partnership whereby one achieves direct benefit from another person's injurious behavior, after the fact, can place the former in silent but unmistakable alliance with what the latter is doing" ([8], p. 10).

Burtchaell's general point, in the form he states it, is hard to disagree with. Benefiting from another's wrongdoing can indeed be part of an alliance with that wrongdoing. But it need not be, as Sulmasy's example of using a murdered person's organs for donation surely demonstrates. More is needed for complicity than benefits after the fact. And Burtchaell recognizes this. His concern is with the legitimation and endorsement of the immoral actions which benefiting from them implies ([8], p. 10).

Burtchaell does not detail the ways in which experimental use of tissue from aborted fetuses would serve to legitimize abortion, except to note how a policy of using aborted fetuses adds a motivation for more abortions ([7], pp. 177-179). But the importance of his argument for bioethical discussion does not require that. For his analysis shows that moral complicity in abortion cannot be limited to causal responsibility for abortions or the desire that they occur.

Sulmasy also argues that the difficulty of informed consent in these cases can be solved. His argument seems to apply to informed consent for transplants from living and from dead aborted fetuses, but since he seems to rule out using living fetuses on other grounds, I will regard his argument as limited to the claim that informed consent for the use of the tissue of a dead fetus can be obtained. He thinks that parents who have consented to abortion are not appropriate proxies for transplant decisions. By this abortion decision they reveal that they do not represent the fetus's best interests, including any the fetus might be thought to have, or to have some duty to have, in contributing to another's welfare. For similar reasons, Sulmasy believes that the state, which permits abortion, cannot be considered a legitimate guardian (cf. also [7], pp. 156-158). But he rejects the position of the most cautious authors, who maintain that no one could be a legitimate proxy. Why not accept Arthur Caplan's suggestion of a committee of third parties, medical staff and other disinterested persons? A body of people having no connection with the abortion is surely not a suspect proxy in the way aborting parents or the state might be ([36], p. 104).

Before Sulmasy's proposal could be accepted as consistent with the cautious approach taken by *Donum vitae*, some difficult questions would have to be answered, for example, about the source of the authority of the committee he envisages. For the fact that normally appropriate proxies are disqualified does not authorize others as proxies just because they are not tainted with what disqualifies the normal proxies. There would also be questions regarding what the proposed committee would deliberate about, and these questions lead naturally to more fundamental questions about the grounds for requiring informed consent for the release of human remains.

Sulmasy assumes a version of a best interests approach, that is, proxies should consider what would be in the deceased person's best interests supposing that these interests are defined so as to include a person's interests in helping others and fulfilling his or her moral obligations. This approach is plausible in cases of proxy consent for therapy, experimentation, and transplantation on behalf of living persons such as children, who have never been able to express their will about such things ([11], p. 30-39). Whatever the difficulties faced by this approach ([27], p. 182-192), its intuitive foundation is powerful: one can hardly respect a person in actions bearing on that person if one does not pay attention to that person's interests, and particularly to those that are morally defined.[3]

But it is not clear that respect for a person who has died implies a concern for that person's interests analogous to that required when the person is alive. Again Burtchaell's account is helpful. He argues that the fundamental ground of the requirement of informed consent for the release of human remains is the need to respect the dead person by treating his or her corpse in ways that reflect the reality of the union between a person and his or her body. We simply cannot separate the way we treat a person and the way we treat that person's remains: "How we treat human remains is both a function and a cause of our bond with human persons" ([8], p. 8). Indeed, the way we treat a person's remains is often the final expression of our respect or contempt for the person. Consequently, we may not simply confiscate or requisition a person's body. For that would discredit the dignity accorded to the person while alive, and treat the person, not just the corpse, as of negligible dignity. Thus, the need for informed consent for the release of human remains.

On this account, a dead person's expressed wishes about the disposal of his or her remains should, if known, be respected. But more than this is plainly required, such as the refusal of actions which will express contempt or disrespect for the person, and, in particular, actions which will establish a new and final expression of disregard for the person's human dignity or which will express acceptance of indignities shown the person while still alive. In other words, on this account, the reasons for requiring informed consent for the release of human remains direct us, in the case at hand, to consider how our using fetal remains may implicate us in the harm the fetus has already suffered by being aborted. This focus might provide Sulmasy's committee a decisive reason not to allow experiments on deliberately aborted fetuses, and it might even give any community opposed to abortion reason to refuse to allow such a committee to exist.

This discussion suggests that *Donum vitae*'s conditions for the legitimate use of the remains of aborted fetuses are most plausibly implemented in the restrictive manner exhibited in the policy Burtchaell defends. Exploring the reasoning involved, however, indicates the need for further thought, particularly concerning the grounds for the commonly held view that human remains can be released only if some appropriate person gives consent.

IV. WITHDRAWING/WITHHOLDING CARE

Catholic bishops and moralists have been active participants in the ongoing American debates about withholding artificially provided food and water from several categories of patients, especially those in a persistent vegetative state (PVS). The treatment of patients in PVS is morally troubling because they are permanently unconscious but are capable of living for a long time if fed, hydrated, and provided nursing care.

There is no official Catholic teaching on the specific conditions under which it is permissible to withhold artificially provided food and water from PVS patients. There is well established Church teaching relevant to decisions of this kind: the prohibitions against suicide and euthanasia, and the doctrine of ordinary and extraordinary treatments ([23]). The most authoritative recent statements are Pius XII's famous 1957 pronouncement on the prolongation of life ([32]), and the "Declaration on Euthanasia" issued in 1980 by the Congregation for the Doctrine of the Faith ([9]).

These Church teachings as articulated in the recent authoritative documents are taken for granted by Catholics who have addressed the issue of withholding food and water from patients in PVS. But this common moral outlook has not led to agreement concerning which specific norms are the correct ones for this class of cases. In this matter, bishops are in public disagreement among themselves and with theologians[4] ([1]; [2]; [13]; [14]; [28]), and there is no theological consensus ([5]; [6]; [16]; [18]; [25]; [26], pp. 380-385; [30]; [31]; [34], pp. 623-647).[5]

According to the doctrine of ordinary and extraordinary treatments, a medical procedure or regimen is extraordinary or disproportionate if it involves an excessive burden for the patient or for others. Treatments which are extraordinary in this way may morally be forgone without any failure to fulfill one's obligations towards preserving life. For in such cases there is no intention of death, and so no possible violation of the Church's prohibitions against suicide or euthanasia, and also no violation of the positive obligations to preserve life, since these are defeasible in relation to the more fundamental spiritual duties whose fulfillment the acceptance of excessive burdens can compromise. By contrast, treatments which involve no burdens or burdens that are insignificant, are ordinary and obligatory. For here there are no grounds for overriding the positive obligations towards preserving life, and at least a suspicion of a wrongful intention towards the patient's life and welfare if such treatments are not provided ([32], pp. 395-397).

Many recent discussions of extraordinary treatments add to this consideration of burdens a further, explicit consideration of whether the treatment does any good. In other words, if a treatment is useless, even if it involves no significant burden, it is considered extraordinary. It is unclear whether uselessness is supposed to be an additional, independent moral factor, or a limiting case of burdensome treatment. But all parties to the debate about withholding food and water from patients in PVS appear to agree that a treatment which is literally useless may be and, indeed, should be forgone.

This brief review of the doctrine of ordinary and extraordinary treatments allows for a statement of the main questions (though by no means all the nuances and subordinate questions) which have emerged in the Catholic discussion of withholding food and water from patients in PVS: (1) Is the withholding of food and water from these patients morally equivalent to intentionally killing them? (2) What are burdens, for the patients or others, involved in artificially providing them food and water? (3) What are the benefits of such treatments, or are they literally useless? (4) If there are burdens and benefits, are the burdens excessive or disproportionate in relation to the benefits?

Since the doctrine of ordinary and extraordinary treatments in effect outlines a procedure for identifying and assessing the morally relevant features of decisions to withhold life sustaining treatments, the existence of some disagreement about specific norms in this area of casuistry is no surprise.

However, the different answers Catholics have given to these questions make clear that the doctrine itself includes significant ambiguities and indeterminacies.

The first question, concerning whether the decision to withhold food and water is intentional killing, is perhaps the easiest to answer, although there is an undercurrent of concern about it which runs through the discussion. Virtually no one in the current Catholic discussion has maintained the strong thesis that withholding food and water from a person who will continue to live if provided them *necessarily* involves the intent of that person's death (cf. [18], p. 298; [26], pp. 381-382).

However, there is no doubt that one can withhold food and water precisely to end or shorten someone's life, and this is intentional killing on the Catholic conception ([16], p. 31; [9], pp. 155-156). Moreover, the statements by some of those involved in decisions to withhold food and water from patients in PVS surely suggest that this intention is sometimes present ([5], pp. 61-62; [28], p. 316). So the worry about intentional killing and euthanasia is present, and it may be more pervasive than explicit discussions suggest.

There is some discussion of the cause of death in patients in PVS. Those who think it wrong to withhold food and water from patients in PVS tend to emphasize that patients in this condition are not dying, and will continue to live if provided nutrition, hydration, and other necessities of life. Thus they maintain that withholding the food and water is the cause of the patient's death ([18], p. 298; [28], p. 315). Those who think it permissible to withhold food and water from patients in PVS reject each of these claims. What causes the patient's death is not withholding the nutrition and hydration but the underlying pathology. Kevin O'Rourke, the chief proponent of this view, seems to hold that once such a pathology is in place, there is no possibility of intending a person's death in a decision to omit treatment, since that context defines the moral issue as a question of whether or not there is a positive obligation to extend life ([29], p. 33; [30], pp. 184-190; [31], p. 352; cf. [2], p. 54).

This particular controversy surely looks like a set of confusions and semantic tangles. For intentions seem more contingently related to the context defined by a patient's condition than O'Rourke allows, and to a decision to withhold treatment than some of the more cautious appear to think; and notions like "dying" and "cause of death" plainly have more than a single meaning. These have yet to be untangled.

The second question, concerning the burdens of feeding and hydrating patients in PVS, reveals one of the deep unsettled questions about the doctrine of ordinary and extraordinary treatments. For those who regard withholding food and water from patients in PVS as wrong tend to limit the burdens of the treatment to one category: the costs, broadly considered, of feeding and hydrating these patients. The normal, patient-centered burdens of medical treatment such as pain, restriction of activities, and so on, do not apply, since these patients are presumed to be unconscious; and there appear to be no other harms or wrongs to the patient involved in these procedures. Other burdens are excluded as improper considerations: the distress of the family is too closely connected with their loved one's tragic situation, and is mitigated only by the end of patient's life; and the sympathetic thought that one would rather die than be in this condition, or that living like this is too great an indignity to bear, is also much more a problem of the person's condition than anything burdensome about feeding or hydrating the person ([5], pp. 62-63; [16], pp. 34-35). But these are burdens which some think appropriate moral considerations ([26], pp. 384-385; [34], pp. 638-640). Plainly, the kind of thing which can count as a burden in the doctrine of ordinary and extraordinary treatments needs clarification.

The third question, concerning the benefits or lack of them in keeping patients in PVS alive by artificial nutrition and hydration, reveals deep disagreements about the nature of benefits. O'Rourke maintains that the doctrine of ordinary and extraordinary treatments rests on the idea that medical interventions should serve the spiritual purposes of life. Permanently unconscious persons, however, can no longer pursue any spiritual goal, and so any medical treatment for such people is useless ([6], pp. 72-73; [29], p. 32; [31]; [32]; cf. [26], p. 383).

Others reject O'Rourke's reading of the tradition ([25], p. 86), and argue that preserving the life of a permanently unconscious person is, as such, a good thing to do. They recognize that this claim is counter-intuitive to many, but argue that denying it amounts to accepting a form of mind/body dualism incompatible with the Catholic moral tradition. The life of a permanently unconscious person is surely deprived, but it remains human life, and thus good ([1], p. 48; [5], pp. 64-65; [16], pp. 36-37; [18], pp. 298-299).

Again, clarification of one of the key notions in the doctrine of ordinary and extraordinary treatments is needed.

Finally, there is the question of comparing the benefits and burdens of feeding and hydrating patients in PVS. There is no generally accepted account of how this balancing is to be done, and the issue is moot for those who think that feeding and hydrating patients in PVS is useless. But some who oppose withholding food and water from these patients have made an interesting argument that the burden is not excessive: the costs of feeding and hydrating these patients is not great, either in comparison to their overall care or to health care costs generally. Moreover, the overall costs of care for PVS patients over time are significant. Thus, it seems reasonable that if costs are what people want to avoid, it is the overall costs, and not the specific costs of feeding and hydrating, that they think are too high. But if the overall costs of treatment are the burden one wishes to avoid, the level at which the burden becomes disproportionate is very high. For avoiding this burden is refusing care, which is tantamount to abandoning the patient. That can be morally acceptable for people in exigent circumstances, but is generally difficult to justify, and for wealthy societies sets a dangerously unjust precedent ([5], pp. 65-68; [16], pp.

32-37).

This consideration of abandonment is a relatively new factor in the discussion; it has not as yet received serious critical attention. Still, it cannot be definitively addressed unless there is further clarification of the general conditions under which the burdens of treatment become excessive or disproportionate, as well as answers to the earlier questions about allowable burdens and benefits.

St. Michael's College
Centre for Bioethics
University of Toronto
Toronto, Canada

NOTES

[1] For useful summaries see [4]; [19], pp. 10-13; [20]; [33], pp. 54-65.

[2] For a different interpretation of this text and a discussion of other official Catholic statements on the beginning of life, see [12], pp. 57-64.

[3] This last point has an implication for *Donum vitae*'s apparently absolute rejection of experimentation on embryos and fetuses, which perhaps should be noted. If, as seems plausible, there are reasons to consent to certain experiments which, if they could be understood and acted upon by children, would be seen by them as morally compelling reasons, and which would be understood by their parents or guardians as imposing obligations on them to consent to the experiments on their child's behalf, then experimentation on children cannot be altogether ruled out. Yet *Donum vitae* is insistent that embryos and fetuses are just like other children. So, there appear to be grounds for qualifying the absolute character of its rejection of experiments on the unborn.

[4] The works cited in the text both report and exhibit this lack of consensus. They contain reference to a considerable body of literature which reveals the development throughout the 1980s of the Catholic debate about withholding food and fluids from patients in PVS. This debate is both historically and conceptually continuous with the wider Catholic discussion of withholding life-sustaining treatments. The current focus on withholding food and water from patients in PVS appears to have displaced the discussion of withholding lifesaving treatments from handicapped newborns as the focus of this wider discussion. See Sparks [35] for a useful critical summary of the discussion of withholding treatments from handicapped newborns and of central aspects of the wider discussion of withholding treatment.

[5] Shannon and Walter ([34], pp. 624-632) confirm the extent of this disagreement in their very helpful report of a survey of the policies of American dioceses concerning withholding food and water.

BIBLIOGRAPHY

1. Bishops of Florida: 1989, "Treatment of Dying Patients", *Origins*, 19, 47-48.
2. Bishops of Texas: 1990, "On Withdrawing Artificial Nutrition and Hydration", *Origins* 20, 53-55.
3. Bone, E.: 1990, "Les Greffes Intracerebrales de Tissu Foetal", *Revue theologique de Louvain* 21, 311-328.
4. Boyle, J.: 1988, "An Introduction to the Vatican Instruction on Reproductive Technologies", *Linacre Quarterly* 55:1, 20-28.
5. Boyle, J.: 1989, "Does the Benefit Outweigh the Burden in the Artificial Provision of Nutrition and Hydration? Affirmative Position", in R. Smith (ed.), *Critical Issues in Contemporary Health Care*, The Pope John Center, Braintree, Massachusetts, pp. 60-69.
6. Boyle, P.: 1989, "Does the Benefit Outweigh the Burden in the Artificial Provision of Nutrition and Hydration? Negative Position", in R. Smith (ed.), *Critical Issues in Contemporary Health Care*, The Pope John Center, Braintree, Massachusetts, pp. 70-77.
7. Burtchaell, J.: 1989, *The Giving and Taking of Life*, University of Notre Dame Press, Notre Dame.
8. Burtchaell, J.: 1988, "University Policy on Experimental Use of Aborted Fetal Tissue", *IRB: A Review of Human Subjects* Research 10:4, 7-11.
9. Congregation for the Doctrine of the Faith: 1980, "Declaration on Euthanasia", *Origins* 10, 154-157.
10. Congregation for the Doctrine of the Faith: 1987, "Instruction on Respect for Human Life in Its Origin and on the Dignity of Procreation", *Origins* 16, 697-711.
11. DiIanni, A.: 1989, "Volunteering Children for Medical Experiments", *Linacre Quarterly* 56:1, 30-38.
12. Ford, N.: 1988, *When Did I Begin? Conception of the Human Individual in History, Philosophy and Science*, Cambridge University Press, Cambridge.
13. Gelineau, L.: 1988a, "On Removing Nutrition and Water From Comatose Woman", *Origins* 17, 545, 547.
14. Gelineau, L.: 1988b, "Statement After Judge's Decision on Medical Case", *Origins* 18, 340-341.
15. Grisez, G.: 1990, "When Do People Begin?", in L. Shrenk (ed.), *The Ethics of Having Children: Proceedings of the American Catholic Philosophical Association* 63, 27-47.
16. Grisez, G.: 1990, "Should Nutrition and Hydration Be Provided to Permanently Comatose and Other Mentally Disabled Patients?", *Linacre Quarterly* 57:2, 30-43.
17. Grisez, G., *et al.*: 1988, "'Every Marital Act Ought to Be Open to New Life': Toward A Clearer Understanding", *The Thomist* 52, 365-426.

18. Grondelski, J.: 1989, "Removal of Artificially Supplied Nutrition and Hydration: A Moral Analysis", *Irish Theological* Quarterly 55, 291-302.
19. Hannon, P.: 1989, "In Vitro Fertilization", *Irish Theological Quarterly* 55, 7-17.
20. Johnstone, B.: 1988, "The Instruction *Donum vitae* and Its Reception", *Studia Moralia* 26, 209-229.
21. Kiely, B.: 1989, "Contraception, In Vitro Fertilization and the Principle of Inseparability", *Linacre Quarterly* 56:2, 68-75.
22. Marcus, E., *et al.*: 1987, "European Reactions to the Document", *Hospital Progress*, July/August, 56-65.
23. May, W.: 1988, "The Simple Case of In Vitro Fertilization and Embryo Transfer", *Linacre Quarterly* 55:1, 29-36.
24. May, W.: 1990, "Catholic Teaching on the Laboratory Generation of Human Life", in M. Wallace and T. Hilgers (eds.), *The Gift of Life: The Proceedings of a National Conference on the Vatican Instruction on Reproductive Ethics and Technology*, Pope Paul Institute Press, Omaha, pp. 77-92.
25. May, W.: 1990, "Criteria for Withholding or Withdrawing Treatment", *Linacre Quarterly* 57:3, 81-90.
26. McCormick, R.: 1989, *The Critical Calling: Reflections on Moral Dilemmas Since Vatican II*, Georgetown University Press, Washington, D.C.
27. McDonnell, K.: 1990, "Volunteering Children", in L. Shrenk (ed.), *The Ethics of Having Children: Proceedings of the American Catholic Philosophical Association* 63, 182-192.
28. McHugh, J.: 1989, "Artificially Assisted Nutrition and Hydration", *Origins* 19, 314-316.
29. O'Rourke, K.: 1988, "Evolution of Church Teaching on Prolonging Life", *Health Progress* 8:1, 29-35.
30. O'Rourke, K.: 1989, "Should Nutrition and Hydration be Provided to Permanently Unconscious and Other Mentally Disabled Persons?", *Issues in Law and Medicine* 5, 181-196.
31. O'Rourke, K.: 1989, "Father Kevin O'Rourke on Hydration and Nutrition: An Open Letter to Bishop McHugh", *Origins* 19, 351-352.
32. Pius XII: 1958, "The Prolongation of Life", *The Pope Speaks* 4, 393-398, also in *Acta Apostolicae Sedis* 49 (November 24, 1957), 1027-1033.
33. Shannon, T., and L. Cahill: 1988, *Religion and Artificial Reproduction*, Crossroad, New York.
34. Shannon, T., and J. Walter: 1988, "The PVS Patient and the Forgoing/Withholding of Medical Nutrition and Hydration", *Theological Studies* 49, 623-647.
35. Sparks, R.: 1988, *To Treat or Not to Treat? Bioethics and the Handicapped Newborn*, Paulist Press, Mahwah.
36. Sulmasy, D: 1989, "By Whose Authority? Emerging Issues in Medical Ethics", *Theological Studies* 50, 95-119.

37. Sutton, A.: 1990, *Prenatal Diagnosis: Confronting the Ethical Issues*, The Linacre Centre, London.
38. Tonti-Filippini, N.: 1989, "A Critical Note", *Linacre Quarterly* 56:3, 36-50.
39. Vacek, E.: 1988, "Notes on Moral Theology: Vatican Instruction on Reproductive Technology", *Theological Studies* 49, 110-131.

COURTNEY S. CAMPBELL

SOUNDS OF SILENCE: THE LATTER-DAY SAINTS AND MEDICAL ETHICS

Although The Church of Jesus Christ of Latter-day Saints (LDS) has a long tradition of reflection on questions of practical ethics, it was not until the mid-1970s that formalized ecclesiastical policy was proposed and adopted on ethical issues arising out of the intersection of religious convictions, medical practice, and applied biomedical technology. First promulgated in 1974 as "Attitudes of the Church of Jesus Christ of Latter-day Saints Toward Certain Medical Problems" [5] and subsequently incorporated in 1976 into the Church's authoritative *General Handbook of Instructions* (hereafter *Instructions*), these initial documents addressed such "problems" as abortion, artificial insemination, contraception, cremation, experimentation, hypnosis, organ transplants, prolongation of life and the right to die, sterility tests, and sterilization. The conclusions presented in these short statements of general policy often did not identify or elaborate underlying theological rationales or ethical principles, or indicate, with some exceptions, as in the case of abortion, how binding these positions were on the practical conduct of Latter-day Saints.

In the intervening years these policies have often undergone modifications in form, sometimes in topic, and less frequently in substance. Nevertheless, the most recent (March 1989) edition of the *Instructions* [10] contains, according to Elder Dallin H. Oaks, the ecclesiastical leader with principal responsibility for revising past policies into their current form, several statements on medical ethics issues "that are entirely new or substantially revised" [15]. This essay will focus on developments in LDS ecclesiastical policy in the areas of reproductive technology, contraception and abortion, termination of medical treatment, suicide and euthanasia, and organ transplantation, as revealed by the 1989 *Instructions*, in which, according to Elder Oaks, "the silent places are as important as its expressed text". Some attention will also be devoted to the very few recent (1985-90) non-ecclesiastical discussions of medical ethics. We shall then conclude with a brief overview of LDS perspectives on clinical professional ethics and on social policy.

A responsible account necessitates some preliminary remarks about the sources and the authority of moral teaching in the LDS Church. The LDS written canon of "standard works" is comprised of *The Holy Bible* (King James Version), *The Book of Mormon*, *The Doctrine and Covenants*, and *The Pearl of Great Price*, and each of these provides important theological principles for LDS moral teaching. Yet, the presence of an "open" canon of scripture illustrates a fundamental feature of the LDS tradition: a conviction of *ongoing revelation* from God to human beings. The "word of God" is thus conceived of as not only the written word of the past but also the spoken or written word of the present. Revelation, including moral teaching, for the Church as a collective community is considered conveyed to a general ecclesiastical leadership, constituted by a First Presidency, the Council of the Twelve, and the Quorums of the Seventy, and while typically not itself canonized, is nevertheless invested with substantial authoritativeness, and may be formalized in the *Instructions*. The constraints of speaking to a broad audience, however, limit ecclesiastical moral discourse to statements of general perspectives that may function as prima facie guidelines, but are silent as to the particulars of moral perplexities.

At the same time, individual members of the LDS Church are understood to be capable of receiving divine direction or revelation for their personal "stewardships" – the practical problems encountered in daily life, in relations with family members, or with other persons for whom they may have pastoral responsibilities. The convergence of the authority and generality of ecclesiastical teaching and the concept of individual revelation entails that the LDS tradition places enormous stress on personal moral agency and accountability in confronting ethical questions that arise in medicine. According to Elder Oaks, "the matters not treated in the [*Instructions*]...are handled by urging the parties to pray and seek spiritual guidance on their own in each individual circumstance" [15]. As we shall shortly discover, that recommendation is in the main applicable as well to the issues that are addressed in the *Instructions*.

REPRODUCTIVE TECHNOLOGY

Three major issues in reproductive technology – artificial insemination, in vitro fertilization, and surrogate motherhood – have been addressed in formal and informal ecclesiastical discourse. Perhaps not surprisingly, given the increasing access to these technologies in our culture within the last decade, their availability has occasioned either reformulation of statements of guidelines previously issued, in the case of artificial insemination, or the first promulgation of guidelines, as with IVF and surrogate motherhood, in the 1989 revisions to the *Instructions*. It is important to identify some underlying theological themes that shape the LDS response to the new technologies of reproduction before describing and commenting on the ecclesiastical guidelines. It will be seen that some of these themes can be at cross purposes in the context of reproductive technology.

1. *The Metaphysics of Embodiment*. In LDS soteriology, each human being lives in an individuated and unembodied spiritual form in a pre-earthly or "pre-mortal" existence. Birth is soteriologically significant, because it provides one with the possibility of obtaining a physical body; embodiment is a necessary, though not sufficient, condition for salvation and eternal life in the presence of God and Jesus Christ, and the body itself is vested with sanctity as a "temple of God". This theological stress on embodiment as a condition of

salvation would seem to support an approving attitude regarding techniques that would enhance the prospects of birth, at least within an LDS family. As Brigham Young, the famous LDS prophet of the mid-1800s put it, "I have told you many times that there are multitudes of pure and holy spirits waiting to take tabernacles, now what is our duty? – to prepare tabernacles for them. It is the duty of every *righteous* man and woman to prepare tabernacles for all the spirits they can..."([13], p. 85).

2. *The Procreative Imperative*. LDS leaders have historically noted that the first divine commandment to human beings recorded in *Genesis* is the instruction to "multiply and replenish the earth", and have continually reiterated that such a command has not been rescinded, even in the face of the world's burgeoning population. While, as we shall later see, the procreative imperative has not meant an absolute prohibition of contraceptive measures, it has, together with the theology of embodiment, reinforced the significance of having children. Once again, reproductive technology might be seen as a positive development consistent with this imperative for infertile couples.

3. *The Sanctity of Family*. The family is an eternal unit in LDS theology, and any action regarding reproduction and contraception will ultimately be justified or limited depending on its impact for family well-being. The value of family also means that it provides the appropriate and desired context and relations for conceiving and nurturing children. As we shall see, it is these features of LDS theology that bear most critically on access to and use of reproductive technologies.

4. *Freedom of Choice*. The "principle of free agency" is considered by the LDS tradition to be inviolable, even by God. This does not of course entail that all choices are morally or theologically equal, but it does mean that in instances of the creation of life, (as contrasted with choices to end life, for example, abortion or active euthanasia), ecclesiastical prescriptions and control may be limited, and direction for decisions may be deemed the responsibility of the principal parties involved, that is, family members. This is in keeping with a commonly invoked standard in LDS discourse that the responsibility of ecclesiastical leaders is to teach correct principles, which in turn allow adherents to govern themselves. With these four themes as theological background, we can now address directly LDS positions on issues in reproductive technology.

Artificial Insemination (AI)

The 1989 *Instructions*, consistent with ecclesiastical teaching since the 1950s, differentiates between artificial insemination by husband (AIH) and artificial insemination by donor (AID). There is no specific stipulation actually provided on AIH, but from both the textual silence and the contrast with AID, it can be assumed that AIH falls into a category of the morally permissible. AID, by contrast, "is discouraged", on the grounds that such a practice "may seriously disrupt family harmony". Even the ecclesiastical opposition to AID has undergone some softening over time; "discourage" is less restrictive than the pre-1977 "does not approve". The prospect of flexibility is further reinforced by the referral of ultimate decisional responsibility for AID to the discretion of the infertile couple: "However, this is a personal matter that ultimately must be left to the husband and wife, with the responsibility for the decision resting solely upon them" ([10], Sec. 11, p. 4).

It is also of significance that the policy on artificial insemination does not differentiate between children conceived through AIH and AID as to their soteriological status. Any such child is considered "sealed" to the parents (the duration of the family bond is eternal), so long as the parents themselves have made a covenant for an eternal relationship within an LDS temple.

This stance of the permissibility of AIH and the discouragement of, but personal responsibility for, AID presupposes that the practice occurs within the context of the family and marital relations. The picture is very different outside this context. The specter of biological relations with multiple anonymous strangers is a likely rationale behind the position that "[t]he donation of sperm is discouraged" ([10], Sec. 11, p. 4). Moreover, access to reproductive technologies by nonmarried women is fundamentally proscribed. In an address at the October 1985 LDS General Conference, the first public ecclesiastical comment on this issue was delivered by Gordon B. Hinckley, the current penultimate "general authority": "I recognize that there are many unmarried women who long to have a child. Some think of bringing this about by artificial impregnation. This the Church strongly discourages. Those who do so may expect to be disciplined by the Church. A child so conceived and born cannot be sealed [eternally united] to one parent. This procedure frustrates the eternal family plan" ([12], p. 89]. Hinckley's comments are essentially reiterated in the 1989 *Instructions*, though even more restrictive wording is used, with "the Church strongly discourages" modified to "is not approved". Finally, because in LDS theology homosexual and lesbian relations are considered violations of the divine standard of sexual morality, it is not surprising to find that the question of access to artificial insemination for gay couples is not even broached.

In Vitro Fertilization (IVF)

LDS Church policy on IVF, which was first articulated in 1989, for the most part parallels the perspective on artificial insemination. Resort to IVF where the gametes are obtained from parties outside the marriage is "strongly discouraged". The term "strongly", which does not appear in the statement on artificial insemination, suggests that there is a distinct and graver moral issue at stake with IVF. While IVF might well be considered a moral notch above AI because of the prospect of creating embryos, freezing embryos, research on or discarding spare embryos, as well as selective termination of multiple pregnancy, the policy does not allude to these issues or elaborate on what might occasion the "strong" discouragement of third-party IVF. Nevertheless, the ecclesiastical flexibility evident in AI is also observed with respect to IVF, for the question of whether to use IVF is designated "a personal matter...[for] the judgment of husband and wife", and the soteriological status of the child conceived through IVF is not conditioned by the source of the gametes ([10], Sec. 11, p. 4).

Surrogate Motherhood

Given the enormous public attention devoted to surrogacy in recent years, the LDS policy on this issue seems somewhat understated: "Surrogate motherhood is discouraged. It might cause spiritual, emotional, and other difficulties" ([10], Sec. 11, p. 5). Indeed, LDS physician and medical historian Lester Bush has noted the "surprisingly moderate" nature of the position on surrogacy [4]. It is less stringent than the "strong" discouragement for IVF, nor does it reiterate the prohibition of access to reproductive technology for single women (though in a parenthetical notation, the *Instructions* suggests that the policies formulated for AI are applicable to surrogacy). No distinction is evidently made between voluntary and commercial forms of surrogacy. Moreover, in this case, soteriology may be conditioned by biology, for under current ecclesiastical policy, a child born through surrogacy would be sealed for eternity to its biological/gestational mother rather than the social or rearing mother [4]. If it is indeed the case that "the silent places are as important as [the] expressed text", the theological and moral anomalies of surrogacy seem to speak volumes on the need for a more comprehensive ecclesiastical statement.

CONTRACEPTION AND ABORTION

The theological principles that establish a strong soteriological and scriptural presumption favoring procreation also work to proscribe substantially methods to prevent procreation, including contraception, sterilization, and abortion. The first word on contraception within the LDS context has historically been the procreative imperative found in *Genesis*, coupled with recommendations for "self-control" in marital relationships. At the same time, ecclesiastical teaching has made accommodations for "the mother's health and strength" and, in some statements, concerns about the genetic legacy of parents, valid reasons to curtail procreation. Moreover, ecclesiastical discussion subsequent to the availability of oral contraceptives has generally affirmed the ultimate decisional authority and responsibility of married couples on this issue.

Some symbolically significant developments in this general tradition have occurred recently. Whereas earlier policy statements on "birth control" had begun by citing the authority of scripture – "The Lord has commanded husbands and wives to multiply and replenish the earth" – this prologue is not retained in the 1989 *Instructions*. In addition, the stipulation of "self-control" has also been excised. The consequence of these changes is a much more liberalized and open-ended policy. Decisions about birth control are exclusively a matter between husband and wife, who "should seek inspiration from the Lord in meeting their marital challenges" ([10], Sec. 11, p. 4). The substantive source for these decisions is now no longer scripture, but personal revelation, which entails that the ecclesiastical directives simply cannot carry the authoritative weight of previous pronouncements.

It should be noted, however, that not all methods of birth control are a matter of ecclesiastical indifference. In particular, significant restrictions are imposed on *sterilization*, which are of some importance, given that sterilization is currently the most popular form of birth control in the United States. As a matter of ecclesiastical guidance, sterilization for purposes of birth control is "seriously deplored". The 1989 *Instructions* generally follows past ecclesiastical teaching in holding that, "surgical sterilization should only be considered (1) where medical conditions seriously jeopardize life or health, or (2) where birth defects or serious trauma have rendered a person mentally incompetent and not responsible for his or her actions" ([10], Sec. 11, p. 5). However, it is no longer sufficient for such decisions to be permissible legally and to be based on "competent medical judgment". In addition, any decision for sterilization is subject to a procedural requirement of consultation and review with local ecclesiastical leaders, as well as "divine confirmation through prayer".

The presence of an ecclesiastical hand is even more prominent in decisions about abortion. LDS ecclesiastical teaching has repeatedly affirmed strenuous opposition to abortion, with some exceptions made for admittedly rare cases, as noted below. However, the prohibition of abortion has posed fundamental perplexities for the tradition regarding the theological or moral rationale for the prohibition and its scope, some of which are reflected in recent changes in official policy. Unlike many other traditions, no definitive LDS position has ever been articulated and formalized on the ontological status of the fetus. Put more theologically, although there is support in LDS scripture for seeing the human person as a soul comprised of *both* spirit and body, no doctrine has been formulated establishing *when* in gestational development the spirit and body form a unified whole. In LDS teaching, then, the ground for the prohibition of abortion cannot be found in a claim that conception, for example, is the decisive moral threshold for the evaluation of prenatal life.

While LDS discourse has frequently seen abortion as symbolic of a general spirit of sexual libertinism, and to that extent, a "revolting and sinful practice" of our times, the most common themes invoked to justify the ecclesiastical opposition to abortion have, ontological agnosticism notwithstanding, portrayed abortion as constituting a specific violation of the general prohibition of killing, or of the principle of reverence for life. The 1989 *Instructions* reiterates the LDS canonical statement, "Thou shalt not steal; neither commit adultery, nor kill, nor do anything like unto it", with abortion being construed as "like unto" unjustifiable killing, or the "shedding of innocent human blood" ([7], p. 108; [10], Sec. 11, p. 4).

This tension between abortion as similar to, but not identical with, unjustifiable killing, is played out in the ecclesiastical sanctions deemed appropriate to abortion. Disciplinary action, including excommunication, may be applied to any Latter-day Saint who encourages, performs, or submits to an abortion; the scope of the sanction thus applies to family members and physicians as well as to pregnant women. At the same time, those disciplined may undergo a period and process of repentance and receive forgiveness from ecclesiastical authorities, a practice that presupposes a distinction between abortion and the taking of innocent human life.

The discourse of ecclesiastical leaders on abortion, including the language of the *Instructions*, suggests that the prohibition of abortion in the LDS tradition should be understood as *absolute*, with certain "exceptions" encompassed by the prohibition out of "sympathetic concern", rather than as a prima facie prohibition that might nevertheless accommodate "justified" abortions. However, since *Roe v. Wade*, which prompted a very emphatic statement of opposition, the range and threshold of exceptions has expanded, with significant additions occurring in 1983 and 1989. Here I will identify the three currently specified "exceptions" and provide some brief historical context:

1. *Life or Health of the Pregnant Woman*. LDS policy as of March 1989 permits an exception to the prohibition of abortion when "[t]he life or health of the woman is in jeopardy in the opinion of competent medical authority". This medical indications policy has remained relatively constant since *Roe*, with some minor modifications: The term "woman" has replaced the earlier term "mother"; "in jeopardy" has been substituted for "seriously endangered"; and "good" has been dropped as a modifier of "health". The substantive significance of these changes is for the most part minimal, though they may suggest an overall lower normative threshold, i.e., "health...in jeopardy" may require less of a burden of proof than "good health...seriously endangered". No differentiation is made in formal policy as to whether "health" should be given a narrow (physiological) or broad (psychological and physiological) interpretation, though the context of earlier policy statements, as well as the ongoing procedural stipulation that indications of life or health as reasons for abortion must be confirmed by medical professionals, may tilt the balance towards a more objective, physiological understanding of "health". Here again, however, the silence of the test on this point leaves open the possibility of practical flexibility.

2. *Violations of the Pregnant Woman*. For LDS Church members, abortions may now be performed when "[p]regnancy has resulted from incest or rape". This exception is substantially broader than statements issued during the mid-1970s in several important respects. First, earlier language had coupled with the rape exception a stipulation that the assault produce "serious emotional trauma in the mother", suggesting that the exception might not apply if the

woman had somehow been able to cope successfully with her victimization. In addition, the late 1970s witnessed some controversy over whether any exception for rape should be allowed, with some leading authorities recommending adoption rather than abortion in such cases so that the unborn child would not be punished for the sins of the rapist. Such a view never prevailed officially, however, and current policy adds no provisos to the rape exception. Finally, the exception for incest was added to ecclesiastical policy as recently as 1983.

3. *Fetal Indications*. In 1989, an exception for abortion was added in situations where "[t]he fetus is known, by competent medical authority, to have severe defects that will not allow the baby to survive beyond birth". Although the exception currently stated is rather circumscribed in relation to similar positions in other religious or professional traditions, the shift in policy should nevertheless not be underestimated. In 1985, for example, Russell M. Nelson, a physician and member of the Church's governing ecclesiastical body, suggested to an assembled general conference that such abortions were performed for "reasons of convenience" and exemplified an "irreverence for life" ([14], pp. 12-13). Perhaps this substantial development reflects awareness of the growing routinization of medical tests in prenatal care, such as amniocentesis, chorionic villus sampling, ultrasound, etc., and their success in detecting abnormalities. At this writing, the LDS tradition had issued no formal guidelines with respect to prenatal screening; silence on these controversial practices may suggest again an ecclesiastical deference to personal responsibility.

The ecclesiastical position on abortion, then, is somewhat less restrictive than at the time of the *Roe* decision, with some important recent changes in each of the above areas, but it is hardly in danger of reflecting pro-choice attitudes. Abortion in the absence of these exceptional circumstances is morally blameworthy and may occasion application of the most severe disciplinary sanctions in the ecclesiastical polity. And, *even* in these exceptional cases, resort to abortion is not a foregone conclusion; there are, instead, significant procedural requirements that must be followed. "The couple" considering abortion must first consult "each other" and "their bishop" (the local ecclesiastical authority), and following a decision for abortion, "receiv[e] divine confirmation through prayer". When one recalls the influential, if not decisive, role of a "competent medical authority" in assessments of a pregnant woman's or fetus's prognosis, it becomes clear that the normative model for a decision about abortion within the LDS tradition is that of *collaborative* choice, rather than an appeal to individual autonomy or privacy. Procedurally as well as substantively, abortion will be a last resort.

A non-ecclesiastical perspective on abortion has been advanced in recent years by a prominent "liberal" LDS scholar, Lowell L. Bennion ([2], pp. 62-64). While for the most part Bennion's conclusions are similar to those of ecclesiastical policy, some important differences can be observed with respect to the moral principles invoked, and their source, weight, and scope. Rather than applying a moral hermeneutic based on the prohibition of killing in scripture, Bennion's analysis draws on the principle of "reverence for life" formulated in the philosophy of Albert Schweitzer. This principle sets a prima facie, rather than absolute, prohibition on abortion, for Bennion holds that abortion is "justified" in several "medical situations". These include the "exceptional" cases recognized in ecclesiastical policy, such as preservation of the mother's life or health and when pregnancy is the result of rape or incest, as well as *in utero* fetal death.

The major substantive divergence is over the justification and scope of abortions for fetal defects. Bennion designates his position a "middle ground" on abortion, in large part because at the time of his writing, there was no officially recognized exception for any abortions for fetal indications. Moreover, while the evolving ecclesiastical policy, as noted above, now draws the line at "*survival* beyond birth", which reflects a sanctity-of-life perspective, Bennion's position incorporates quality-of-life considerations, in that abortion can be justified "when the child has such severe mental and/or physical handicaps that no *meaningful life* he or she will experience is possible" (my emphasis). However, Bennion does not elaborate on how the qualitative criterion of "meaningful life" is generated out of the principle of reverence for life, or what kinds of life experience would qualify as lacking meaning. In an important sense, the emphasis on embodiment in the LDS tradition entails that any post-birth experience, no matter how short, severe, or debilitated, has ultimate and eternal soteriological meaning.

TERMINATION OF TREATMENT

The capacity for current medical technologies to extend life almost indefinitely has posed for the LDS tradition, no less than for other religious and professional traditions, perplexities and dilemmas about when, if ever, it is appropriate or justifiable to forgo such measures or permit their cessation. Several themes in LDS theology can be briefly identified that condition discourse within the tradition on the termination of life-prolonging medical treatment.

1. *The Sovereignty of God*. A theological presumption in favor of life is established by the dominion of the Deity over life and death. Life is a fundamental, though not absolute, good; death is neither to be sought nor seen as an ultimate enemy. It is a passage in human progression and may in certain circumstances be viewed as a "blessing". In addition, decisionmakers confronted with end-of-life issues should seek divine guidance and not neglect the prospect of divine intervention in healing.

2. *The Stewardship of Human Beings*. Since human beings are created in God's image, their relationships always presuppose a radical sense of both dependence and equality. There is no fundamental moral discontinuity between the dying person in the bed and the caregiver at the bedside; what they hold in common is more significant than the fact of illness that separates them.

These themes of dependence and equality form the theological parameters of stewardship, in which Latter-day Saints are called on to care for and meet the needs of others, for whom they are ultimately responsible and answerable. The pattern for stewardly responsibility is God, exemplified in the life of self-giving care of Jesus Christ. This pattern, moreover, entails a specific stewardship for the most vulnerable and voiceless in the community, especially the seriously ill.

3. *Moral Agency*. As we have seen previously, the principle of free agency in the LDS tradition requires respect for the self-determined decisions and choices of others. Such choices, particularly in the context of death and dying, need to be informed and voluntarily assumed; a balanced stewardship will forgo the temptation to define the welfare of others by values they do not claim as their own. At the same time, a responsible choice will be the outcome of a collaborative procedure, in which patients, family members, and medical caregivers, but *not* necessarily ecclesiastical authorities, are important participants.

These theological principles form the context for positions in the LDS tradition on the termination of treatment that (a) establish a general presumption in favor of utilizing available medical technology to sustain life, (b) permit forgoing medical interventions in some instances to allow a patient to die, and (c) prohibit assisted suicide and active euthanasia in all situations [6].

Although the formal LDS Church policy on termination of treatment articulated in the 1989 *Instructions* reiterates for the most part the ecclesiastical position adopted since 1976, some modifications seem to provide a more accommodating stance to the withdrawal or withholding of life-prolonging treatment. According to the section "Prolonging Life", in instances of "severe illness..., Church members should exercise faith in the Lord and seek competent medical assistance" ([10], Sec. 11, p. 6). This recommendation is an example of the tradition's "faith *and* works" soteriology carried over to the medical context. The exercise of faith may in part mean asking a Church elder (a lay position held by most adult LDS males) for a "priesthood blessing" or an "administration to the sick", which involves anointing the head of the afflicted person with consecrated oil and invoking a blessing on their behalf, in conformity to the direction given in the New Testament: "Is any sick among you? Let him call for the elders of the church; and let them pray over him, anointing him with oil in the name of the Lord: And the prayer of faith shall save the sick, and the Lord shall raise him up...." (*James* 5:14-15). The exercise of faith might also include prayers by family members, and worship services for medical professionals and caregivers to function as mediators of the divine presence in the healing of the ill. In the broadest sense, then, faith and the life-saving resources of medicine are seen neither as antithetical nor as identical, but as complementary and reciprocally related.

"Prolonging Life" continues to affirm that, "when dying becomes inevitable, it should be looked upon as a blessing and a purposeful part of eternal existence". "Eternal" in the 1989 *Instructions* has replaced "mortality", a modification that has no substantive bearing on the question of terminating treatment, but simply reiterates the LDS conviction that death is but a passage in a life span that both precedes and continues beyond mortality. There is, it should be noted, potential ambiguity in the use of "it" in the above clause; the text is silent on *who* should consider death a "blessing". One can certainly conceive of situations of intrafamilial or family-caregiver dispute over whether dying has become inevitable, occasioning the blessing of death.

Perhaps the major substantive change in the 1989 *Instructions* is the *first* formalization of an "unreasonable means" threshold for forgoing treatment. "Members should not feel obligated to extend mortal life by means that are unreasonable". Again, there is textual silence on what criteria might be used to determine what constitutes "unreasonable", although as indicated below, the strong implication is that resort to a procedural solution would be appropriate. In any case, the striking point to emphasize is that the inclusion of a discretionary "unreasonable means" threshold comes in the same document that omits long-standing language that made it obligatory to ask medical practitioners "to assist in reversing conditions that threaten life". The sum of these two changes presents, if not an endorsement, at least a much stronger affirmation of the permissibility of terminating medical treatment than previous policy statements.

Finally, "Prolonging Life" instructs that "[t]hese judgments are best made by family members after receiving wise and competent medical advice and seeking divine guidance through fasting and prayer". The antecedent of "these judgments" is not the intellectual concern for classifying certain means as "unreasonable" or not, but rather the practical concern about whether or not to prolong treatment. In this collaborative process, of course, it is likely that criteria of the "unreasonable" will be implicitly invoked or assumed, but conceivably *any* medical means that prolongs life, including intravenous nutrition and hydration, could be deemed unreasonable by the participants. Significantly, the ultimate locus of decision-making authority is portrayed as residing within the family. On the one hand, unlike policy on abortion, this context excludes ecclesiastical leaders and, to a certain extent, caregivers, who function more as advisors on the outside than internal participants. On the other hand, it also suggests there is no automatic acquiescence to the wishes of the particular patient; autonomy is here qualified by the relational context. There is, finally, no attention given to legal procedures, such as advance directives or living wills, which might resolve lurking controversies in advance. Indeed, the recommendation to seek "divine guidance" in time of severe illness could be construed to suggest that a form of situationism is preferable to general ecclesiastical directives or prior stipulations in legal instruments. As we shall now see, the situation is rather different regarding questions of suicide and euthanasia.

SUICIDE AND EUTHANASIA

For a substantial period of its history, the LDS tradition affirmed a strong condemnation of suicide. It was deemed self-murder and comparable to the sin of "shedding of innocent blood", meaning in LDS theology that suicide constituted an unforgivable sin that would culminate in the denial of eternal blessings and rewards in the afterlife. The soteriological status of the suicide was reinforced by a kind of disenfranchisement of the deceased; funerals for persons who had taken their own lives were to be private rather than public ceremonies, and were to be performed in the absence, at least in any official capacity, of ecclesiastical leaders.

More recently, however, an explicitly less judgmental tone has been adopted towards the person who commits suicide, so that the person is distanced from the act, and attribution of culpability is deferred to divine determination. Thus, in a revision of previous policy, the 1989 *Instructions* states: "A person who takes his own life may not be responsible for his acts. Only God can judge such a matter". Moreover, official ecclesiastical involvement in funeral services is clearly encouraged: "The family, in consultation with the bishop, determines the place and nature of a funeral service for a person who has died under such circumstances. Church facilities may be used without restriction" ([10], Sec. 11, p. 5).

Outside formal ecclesiastic policy, this more empathetic spirit is exemplified in a short article, "Suicide: Some Things We Know, and Some We Do Not", authored by M. Russell Ballard, a member of the governing ecclesiastical body. Ballard is at pains to reiterate traditional teaching of the objective wrongness of suicide. Suicide is "a very grievous sin", and "truly a tragedy because this single act leaves so many victims: first the one who dies, then the dozens of others – family and friends – who are left behind, some to face years of deep pain and confusion". Nevertheless, Ballard holds out "hope" (for both "victim" and survivors), rather than the prospect of condemnation and eternal punishment, because of uncertainty over the autonomous nature of the act. Ballard raises several circumstances that may mitigate culpability – mental illness, severe depression or emotional disturbance, a call for help, insufficient comprehension of the act's seriousness, or a chemical imbalance. Because "we do not know the full circumstances surrounding every suicide", he contends, judgment of the *person* and his or her eternal destiny is reserved to the omniscience, justice, and perfect love of God [1].

From Ballard's analysis, three principal conclusions can be drawn that seem to reflect the current LDS perspective on suicide. First, though suicide cannot be *justified* morally or theologically, circumstances of a particular act may provide *excusing* conditions or reasons. Second, the theological analysis (admittedly motivated by pastoral concerns for survivors) would call into serious question the philosophical concept of "rational suicide". Although Ballard does not commit himself either way as to *whether* suicide could ever be rational, it is clear that if a case can be made for the rationality of suicide in *particular* cases, the theological differentiation between assessment of the act and the person would collapse. Finally, the operational assumption for the evaluation of suicide reflects a traditional LDS principle: There is no accountability or culpability without agency.

To this point in our survey of LDS attitudes on questions of medical ethics, we have seen on most issues a gradual liberalizing of positions over time. The development of policy on euthanasia proves the primary exception to this; even as (and perhaps because of) societal and professional tolerance and support for active euthanasia increases, the LDS position has become increasingly definitive and rigid. Thus, in 1974, the question of euthanasia was treated as part of a section on the "Prolongation of Life and Right to Die" and assessed negatively: "The Church does not look with favor upon any form of mercy killing". This was subsequently altered to, "[b]ecause of its belief in the dignity of life, the Church opposes euthanasia". By 1989, however, euthanasia rated its own specific section of the *Instructions*, and the evaluation was even more descriptively and normatively definitive: "A person who participates in euthanasia – deliberately putting to death a person suffering from incurable conditions or diseases – violates the commandments of God" ([10], Sec. 11, p. 5). Simply put, the condemnation of euthanasia now is much stronger in that the teaching authority of "the Church" is mediating the divine will, rather than affirming "its belief".

The basis for the prohibition of euthanasia appears to follow rather straightforwardly from some of the above themes. Euthanasia would seem to violate the principle of responsible stewardship and usurp the sovereignty of God over life and death. It does not, moreover, seem to make any moral difference whether euthanasia is voluntarily requested or beneficently motivated; the prohibition of "*any* form of mercy killing" in the 1974 statement is more explicit on the former issue, while the current position does not take into account any intention of compassion or "mercy" in euthanasia, which was at least alluded to previously. In both cases, the moral focus is on act and outcome, not on intention or the autonomous nature of the request. Indeed, in contrast to the analysis of suicide, it is important to observe that the negative assessment of euthanasia is directed at "a person".

Nevertheless, the current position contains some striking silences. Even if euthanasia "violates the commandments of God", it is not exactly evident *which* commandment(s) is (are) violated. The earlier language of "killing" has been replaced by "deliberately putting to death", terms that may simply be interchangeable. Yet, if euthanasia were to be assessed under the concept "killing" (albeit, unjustifiable), one would expect at least some mention of ecclesiastical sanctions and discipline similar to those present in the discussion of abortion. Similarly, if euthanasia constituted a case of the "shedding of innocent blood" in a medical context, some discussion of the social and legal status of the act would be appropriate.

In other instances, textual silence seems an occasion for the assumption of personal responsibility and the withdrawal of ecclesiastical authority. I do not think the same lesson can be drawn from these omissions about euthanasia, however. As incomplete as the current policy may be, its prohibition of the act and condemnation of the person are about as absolute and definitive as LDS ethical teaching ever gets.

ORGAN TRANSPLANTATION

The success of the first solid organ transplants in the 1950s and 1960s occasioned substantial enthusiasm in the medical community and society as a whole over the prospect of prolonging life and achieving deliverance from death through this new "gift of life". However, such anticipatory attitudes were not displayed within the LDS tradition. On the contrary, Lester Bush has recounted that during the 1960s, Latter-day Saints who had participated in rituals of covenant-making and salvation in LDS temples (the "endowment") were counseled not to will their organs for use in medical research, nor to participate as living donors, a principal source of organs at the time. No public rationale was ever offered for these recommendations, although Bush speculates that LDS doctrines of a literal resurrection of the physical body and the sanctity of the body as a temple of God may have motivated ecclesiastical reservations about organ transplantation [4].

The public LDS policy on organ transplantation is best described as nonexistent. A statement in the 1974 "Attitudes" that remained virtually unchanged through the mid-1980s makes the matter rest entirely on personal agency; the decision to "will" (not "donate") organs for purposes of transplantation or research is deemed resolvable only from "deep within the conscience of the individual involved". Potential donors were informed that they would not receive ecclesiastical guidance on the question, but should instead obtain knowledge about the "advantages and disadvantages" of the practice, seek divine direction, and ultimately adopt "the course of action which would give them a feeling of peace and comfort" ([5], p. 98) The "disadvantages" alluded to here presumably could refer to risks to the recipient from a failed procedure, risks to the donor if living, or possibly ultimate risks attributable to the unresolved ownership of transplanted organs in the resurrection. While imbued with individualism, the statement in historical context is certainly in conformity with the spirit of the times, as the Uniform Anatomical Gift Act gave decision-making primacy to the individual, with the primary source of organs viewed as authorization on a personal donor card.

Despite substantial successes in transplantation technology and major policy changes in procurement methods turning the 1980s, revisions on the subject of organ transplantation introduced in the 1989 *Instructions* are relatively minor, and continue to reflect the posture of personal rather than ecclesiastical responsibility. To be sure, the family is now explicitly recognized as a potential decisionmaker in cases where authorization is needed to procure organs from a deceased family member, but the substance of the choice continues to remain a matter of personal or family discretion, with neither encouragement nor discouragement conveyed in the policy. Moreover, the act continues to be designated legalistically rather than morally; a person "wills" his or her organs rather than making a "gift" or "donation" to others. A potential recipient (here the language does shift to a "donated" organ) is encouraged to rely on both "competent medical counsel and confirmation through prayer" in determining whether to undergo a transplantation procedure ([10], Sec. 11, p. 6).

This ecclesiastical non-directiveness has, not surprisingly, made it possible for organ transplantation to obtain a more enthusiastic endorsement by lay members and physicians. U.S. Senator Jake Garn (R-Utah) received national attention in 1986 for undergoing a transplant procedure to provide a kidney for his daughter. Moreover it should be recalled that one of the most innovative biomedical experiments of the last decade, the artificial heart, was first performed on an LDS patient, Barney Clark, by an LDS surgeon, William De Vries. More recently, a short opinion piece that gave strong encouragement to consideration of organ transplantation appeared in the *Ensign*, the official LDS monthly publication, authored by physician Cecil O. Samuelson, Jr. While acknowledging that the decision to participate in organ transplantation remains a personal matter "deserv[ing] prayerful consideration", Samuelson's positive assessment is perhaps best exemplified in his observation that transplantation constitutes a "gift of life and health" that expresses "selfless love". Samuelson also attempts to dispense with the objection to transplantation on grounds of concern about bodily wholeness in the resurrection by appealing to LDS scripture: "It seems obvious ... that organ transplantation does not affect one's resurrection, since the organ would soon have returned to the basic elements of the earth following death anyway. Whatever happens to an organ following death, we are promised that 'every limb and joint shall be restored to its body, yea, even a hair of the head shall not be lost' (*Alma* 40:23)" ([16], p. 50).

Perhaps more striking than this hermeneutical defense of transplantation, however, is the attention Samuelson gives to living, related donors. Participation in a transplantation procedure, he (correctly) observes, has frequently been an important source of consolation for families experiencing grief over the death of a member, while "[o]ther families have been spared debilitating illness or death because a living family member was able to donate an organ to a loved one." Samuelson then proceeds to reassure prospective living donors that transplantation technology and medical screening have both enhanced the prospects of a successful procedure for the recipient and diminished the physiological risk for the donor, making little impact on the donor's lifestyle.

Although there is no concept of supererogatory action in LDS doctrine that might motivate organ donations, it also seems that there are no substantive theological arguments against the practice. Personal agency is the decisive consideration on this issue, and the way is entirely open for Latter-day Saints not only to be donors of organs through societally required request protocols, but also to act as living donors for members of their families.

CLINICAL AND SOCIAL ETHICS

The range of bioethics issues that have received formal ecclesiastical attention or discussion among lay LDS authors is, as illustrated by this overview, rather limited. For the most part, discourse about bioethics within the tradition has focused on issues related to the creation and ending of life, the very issues presupposed in the "playing God" analogy in bioethics. By contrast, such important questions in the clinical practice of medicine as research with human subjects or confidentiality in the patient-health care professional relationship have generated little response. Similarly, vital health policy questions that lie at the intersection of ethics, medicine, and law, such as maternal-fetal conflicts, the setting of health care priorities, and the definition of death, have in large measure gone unaddressed. In this concluding section, I shall suggest that this response of silence is related to fundamental perspectives about the ethical integrity of medicine and about ecclesiological identity, as expressed in a social ethic of an exemplary community.

In the mid-1970s, ecclesiastical guidance of a very general nature was provided on research with human subjects: "The Church recognizes the need for carefully conducted and controlled experimentation to substantiate the efficacy of medicines and procedures. We believe, however, that the free agency of the individual must be protected by informed consent and that a qualified group of peers should review all research to ascertain that it is needed, is appropriately designed, and not harmful to the person involved" ([5], p. 100). The current *Instructions*, however, does not contain any reference to experimental procedures, an omission that may seem surprising given the prominence of LDS participants in the early trials of the artificial heart. Part of the explanation for this might be attributed to the emergence in intervening years of Institutional Review Boards (IRBs) as a procedural mechanism to govern clinical research, including precisely those features – informed consent and the scientific and moral adequacy of a research protocol – that had animated earlier ecclesiastical concern.

This, however, exemplifies a more fundamental point, with broad applicability, about the attitude of the LDS tradition to modern clinical medicine, namely, the presence of a profound respect for the moral integrity of the healing professions. We have, in our survey of issues that have commanded ecclesiastical attention, seen that respect implicit in the frequent recommendation that medical professionals (not ecclesiastical officials) be viewed as primary consultants in cases of medical-moral perplexity. In less controversial issues (i.e., issues in which the creation or ending of life is not the core concern), the LDS tradition seems to see the domain of medicine as ethically self-sufficient and adequately self-regulating. Put another way, the tradition seems to affirm a posture of compartmentalization, rather than the generalization, of expertise.

This respect for the scientific and ethical self-sufficiency of medicine is also suggested by a "pattern" discerned by Lester Bush in LDS ecclesiastical responses to medical ethics. Bush has observed that ecclesiastical guidance is typically retrospective rather than prospective, and that in many instances, no formal guidance is articulated when "the issues are extraordinarily complex or when important scientific questions remain unanswered" ([3], p. 48). No policies have ever been developed, to my knowledge, for issues in genetic engineering, for example. Further, Bush maintains, formal public statements on specific issues typically are not issued until the public debate is on the verge of ending, and they "invariably" evolve in "the direction of greater conformity to the general medical/social consensus on the subject" ([3], p. 49). In a fundamental sense, the ecclesiological identity of the LDS Church is not directed towards solving the problems of biomedical science and their clinical applications.

If there is an ecclesiastical reticence to enter formally into the domain of clinical practice and medical ethics, this is perhaps even more pronounced when the complex "scientific" questions become intertwined with equally difficult matters of law and politics in the realm of health policy. In the last decade, for example, only two "political" issues have occasioned formal ecclesiastical direction and involvement, the Equal Rights Amendment and the MX missile, both of which were opposed.

There are certainly health practices of Latter-day Saints that have implications for health policy issues, especially cost-containment measures. Chief among these is the "Word of Wisdom", which is obligatory for all members in "good standing". This law of health, which has the status of a "commandment" in the tradition, prohibits smoking, drinking alcoholic beverages, coffee and tea, and the use of "any substance that contains illegal drugs or other harmful or habit-forming ingredients" ([7], pp. 175-76; [10], Sec. 11, p. 6). Both epidemiologists and health policy analysts have observed how such practices have produced among Latter-day Saints the lowest rates of mortality and cardiovascular disease of any group in the United States ([8]; [9]). The question of encouraging, promoting, or requiring a particular life-style obviously has bearing on issues of cost-containment and priority-setting in health policy.

The *social* ethic of the LDS tradition must be sought in these practices, not in public discourse or statements; the practices *are* a social ethic of a community that tries to witness to values and alternative patterns of life, including healthy life-styles, rather than feeling compelled to formulate an "LDS" position or policy for discussion in the public square. Some LDS scholars have called recently for more pronounced ecclesiastical involvement in the political forum, asserting that "the Church ought to play the role of moral critic of other institutions of society" and be "actively engaged in political issues that relate to human welfare – particularly those which affect protection of life, freedom, and social justice" ([2], pp. 81-82). However, social activism and advocacy for the most part continues to be rejected as an institutional calling by the ecclesiastical leadership on the grounds that it will "fracture our strength by leading us in the pursuit of objectives which are not pertinent to the central mission of the Church" [11]. So long as such a posture prevails, the LDS tradition will likely confront the emerging health policy issues of contemporary bioethics with ecclesiastical silence and an affirmation of personal and professional responsibility.

Oregon State University
Corvallis, Oregon, U.S.A.

BIBLIOGRAPHY

1. Ballard, M.R.: 1987, "Suicide: Some Things We Know and Some We Do Not", *Ensign* 17:10, 6-9.
2. Bennion, L.: 1988, *Do Justice and Love Mercy: Moral Issues for Mormons*, Canon Press, Centerville, Utah.
3. Bush, L.E., Jr.: 1985, "Ethical Issues in Reproductive Medicine: A Mormon Perspective." *Dialogue* 18, 58-60.
4. Bush, L.E., Jr.: *Health and Medicine Among the Latter-day Saints: Science. Sense and Scripture* forthcoming.
5. Bush, L.E., Jr. (ed.): 1979, "Mormon Medical Ethical Guidelines", *Dialogue* 12, 97-106.
6. Campbell, C.: 1989, "At the Edges of Life", *This People* 10:3, 21-24.
7. *Doctrine and Covenants*: 1986, The Church of Jesus Christ of Latter-Day Saints, Salt Lake City.
8. Enstrom, J.: 1989, "Health Practices and Cancer Mortality Among Active California Mormons", *Journal of the National Cancer Institute* 81:23, 1807-1814.
9. Fuchs, V.: 1974, *Who Shall Live? Health, Economics, and Social Choice*, Basic Books, New York.
10. General Handbook of Instructions, 24th ed.: 1989, The Church of Jesus Christ of Latter-day Saints, Salt Lake City.
11. Hinckley, G.: 1983, "He Slumbers Not, Nor Sleeps", *Ensign* 13:5, 5-8.
12. Hinckley, G.: 1985, "Ten Gifts From the Lord", *Ensign* 15:11, 86-89.
13. McConkie, B.: 1979, *Mormon Doctrine*, 2nd ed., Bookcraft, Salt Lake City.
14. Nelson, R.M.: 1985, "Reverence for Life", *Ensign* 15:5, 11-14.
15. Oaks, D.H.: 1990, personal communications.
16. Samuelson, C.O.: 1988, "I Have A Question", *Ensign* 18:2, 50.

PRAKASH N. DESAI

HINDUISM AND BIOETHICS IN INDIA: A TRADITION IN TRANSITION

There has hardly been a time in Indian history when the Indian tradition has not been in transition, except perhaps at the very beginning of Indian civilization. In coming face to face with the West, India may appear to be a culture wrapped in its past, possessed by religion, and aglow in its mystical tradition. But India speaks in many voices, and this polyphony makes it difficult to articulate a singular judgment. An evocative metaphor often employed to capture the spirit of the tradition is that of a palimpsest, an ancient parchment upon which generation after generation inscribe their messages layer upon layer, and yet no succeeding layer completely erases or hides what was previously written ([56], p. 59).

This paper is concerned with the problems of contemporary medical ethics. The reader will not find clear-cut confrontations between tradition and modern medicine in this essay. Instead, poverty will occupy the main stage; there will be glimpses of internationalization of medicine, values in time warps, and the pervasive influence of what goes on in the rest of the world, ethically and medically, sometimes overriding local systems of values. In order accurately to appreciate the medical tradition and the ethics that flow from the religious mandates, an acquaintance with the evolution of medical practice and the development of the cultural ethos is essential. Approaching problems of medical ethics in India is not straightforward. Neither the medical nor the religious community or their leaders have so far voiced audible concerns about these issues. Mostly the debates go on in newspapers reflecting public opinion and attitudes. In India it is not easy to formulate ethical principles that might uniformly apply to all groups and therefore mass media are an appropriate outlet.

THE BEGINNING

The earliest known inhabitants of India were the people of the Indus Valley civilization, also known as the Harappa civilization from the name of one of its cities which flourished about 2500 B.C.E., and which was excavated only in the present century. The Harappans were the first builders of sewers, sanitary soak pits, and public bathhouses, but were also intensely conservative in their architectural preferences, building identical cities over cities after successive river floodings ([10], p.16). Among the many finds is a seal with a man seated in a yoga position, prompting speculation that the practice of yoga may have originated with the Harappans. They were overwhelmed around 1500 B.C.E.

by herds of Indo-Europeans. Ever since the arrival of these cattle-breeding, pastoral Indo-Aryans, a linguistic rather than an ethnic grouping (and it is difficult to imagine that this was a singular migration and not a series of tribal incursions over centuries), the land and the shores of the Indian subcontinent have known wave after wave of immigrants and invaders: the Greeks, the Shakas, Kushans, and Huns in the early period; the Arabs, Afghans, Turks, and the Moghuls in the middle period; and the Portuguese, French, and the British in the more recent past. Assimilation of immigrants and accommodation of the invaders has given the culture every conceivable new twist at different times and in different places.

The word Hindu is one such consequence. River Sindhu, which now mainly flows through Pakistan, was known to the Greeks as Indus, and to the ancient Persians as Hindu; hence the people that lived around the river were called Hindu, their land India, and their religion Hinduism. The early period of the settlement of the Indo-Aryans was marked not only by conflict with the native civilization, but also by tribal strife and internecine warfare. As the population began to disperse to the east and the south, they contended with many local tribal organizations. What began as a movement away from the center of the Indo-Gangetic plains became a characteristic cultural pattern; centrifugal forces continued to reshape Indian thought and practices. Hinduism, as it came to be known later, was a product of the interaction among the Vedic tradition of the Aryans, the civilized Harappans, and several groups of tribal inhabitants. Hinduism in turn spawned the new religions of Buddhism and Jainism about 500 B.C.E. During the same period, the religion that had no "written word" or prophetic commandments to bind its people developed divergent and sometimes conflicting interpretations about the nature of reality and divinity. These very same forces led to a variety of worship forms and religious practices. Innumerable sects organized themselves, as new ones continue to emerge even today, often around the personage of a guru or a charismatic interpreter of religion. But there was never a full-scale revolution, despite revolts; never a clean break, only significant divergences and slight aberrations.

Perhaps because of repeated insults from incursion and invasion, or an intrinsic capacity of the Hindu tradition to co-opt its revolutionaries, the past remained an evocative motif for continuity.

It is also at this stage that the caste system became firmly established. Social groups became the purveyors and enforcers of codes of conduct. The business of maintaining public morality came to be the province of social groups, namely, the innumerable *jatis* (the word is etymologically related to *genera*), the subcastes, which added to the diversity of ethical norms with their regional character. The caste system has been extensively written about, and here it is sufficient only to mention that each caste has different moral obligations, derived from its constitutions ([52]; [23]). Morality was defined in terms of actions and behaviors appropriate to the social group or sect to which one belonged. *Dharma*, from the Sanskrit *dhru*, meaning "that which holds together" (for a person or a group, each according to its nature), became embodied in the law books, the *Dharmashastra*. Sources of *dharma*, an issue of interminable debates, were regarded to be revealed literature, *shruti*, i.e., "that which was heard", from which was derived the remembered word, *smruti*, the law books being a part of the later. When the population dispersed, different regions and different clans composed a new version of the law books to accommodate to local conditions. The composers of law texts added a twist to the sources of *dharma*: the authority of moral conduct of respectable men ([38] II, pp. 1-10; III, pp. 825-884). This marked a further contextualization of *dharma*, which increasingly meant that moral obligations and interpretations were relativized to local customs. The redactors were sensitive enough to accord contingencies of time, place, and personal circumstances a significant status. The caste system thus produced laws that applied to different groups in different ways. The personal context reached its zenith when the lawmakers decreed that what was pleasing to the self was an acid test of the appropriateness of *dharma*.

HISTORY OF MEDICINE

Practice and patronage of medicine in India is truly pluralistic. Every known medical tradition thrives in India, and people patronize practitioners from different systems, sometimes simultaneously. An organized body of knowledge called *Ayurveda* developed during and after the rise of Buddhism in India, and was rendered in a textual form in the first century C.E.: the *CarakaSamhita*, a compendium attributed to a legendary physician Caraka ([77], pp. 222-263). It is suggested that the word *Caraka* means "one who moves about" and may have been applied to Buddhist monks. Chattopadhyaya argues that the physicians were held in contempt by brahminical orthodoxy because of their pursuit of a materialist science, and because of their "democratic practice of mingling with the common people", regarded by the priestly authorities as "promiscuous", and "unaristocratic" ([17], pp. 246, 241). Thus, from the beginning, there has always been a chasm between religious ethics and the ethics of physicians in practical realities. But the infrastructure of this medicine, i.e., the constructions of the body and the self, the nature of pain and disease, and the logic of therapeutics, derived from the cosmology and ontology of the Vedas, Upanishads, and schools of Samkhya and Yoga ([20], pp. 35-57, 73-112). Over time the two views of life, the medical and the religious, grew in tandem and in mutual interaction. Just as religious thought and practice were evolving in confrontation with tribal organizations and ancient native conceptions, so was medicine. What Marriott ([51], p. 197) describes as universalization and parochialization, processes of mutual penetration of the great tradition and the several local traditions, applied to the upward and downward assimilation in Indian medicine.

With the invaders also came their medicine, and what the rulers patronized also tended to dominate. With the Greeks came their tradition. Arabs brought *Unani* medicine, and the allopathic tradition came with colonizers from the West. Homeopathy, biochemy, and naturopathy have all found a comfortable home and a sizable following in India. Ayurvedic medicine developed further specialization; for example, the *Siddha* tradition, patronized mainly in the south of India, took on the Greek apothecary, especially treatment with metal oxides. The overwhelming impact of the Arab tradition on Ayurvedic practice was the examination of the radial pulse.

Western medicine, referred to as allopathy in India, has, of course, become the most dominant, with over 100 medical colleges, half a million hospital beds and 300,000 medical practitioners. The Ayurvedic tradition has a similar number of practitioners, close to 100 medical colleges, but only 20,000 hospital beds [35]. Health economists should take note: home-based health care is the most popular and viable form of treatment.

CONTEMPORARY MEDICAL REALITIES

Western medicine was introduced by the British and, at the beginning, mainly served the colonial interests of the British administrators and their army. Save for epidemics, the local populations were not the beneficiaries.

The health delivery system as it exists today is an amalgam of the old and the new. Competing and complimentary medical systems exist in India with variable distribution between the urban and rural areas. Without exaggeration, it can be asserted that the ethical crisis in India is the problem of access to health care. Rural areas, especially those farther away from cities, have no access to health care; they can neither afford it, nor is it in easy reach. Although indigenous medical systems have some presence in rural townships, and people travel to them with primitive modes of transportation, they are of no avail in acute medical crises. Between competing traditional and modern medical systems, the folk wisdom has accurately shown preference for the modern in acute illnesses and for the traditional in the more chronic [31]. Despite the image of a socialist orientation, health care in India is mainly for those who can afford it.

There is no national health insurance. In fact, no health insurance industry exists. A few, those who are employed in the government and industrial sectors, have a modicum of coverage, but this is a minuscule population. Primary health care centers, funded by the state and central government, are few and far between, poorly staffed, and often lack a physician. Such medical posts are unattractive, both because of location and low remuneration.

Over the last four decades, in spite of bold local attempts to create human resources for health care, the developed world has siphoned off a large number of physicians and nurses. The British Public Health Service is largely run by physicians from Asia, particularly from the Indian subcontinent, and the same group composes about 5% of the physicians in the U.S. Some view this problem of "brain drain" as a form of foreign aid from the developing countries to the developed world.

In the metropolitan areas of the country, such as Bombay, Delhi, Calcutta, and Madras, modern technology has arrived. A few physicians returning from the West have joined in commercial enterprises to staff ultra-modern hospitals which cater to the very rich. Transplants, coronary by-pass, lithotripsy, scans, and other sophisticated interventions have become available [72].

An explosion in health care delivery did begin after India's independence. The blueprint for the development was the report of the Bhore Committee, established by the British, which submitted its findings on the eve of independence. Among its major recommendations were (1) universal access to health care, (2) primary prevention, (3) emphasis on rural areas, and (4) training physicians to be socially oriented ([9], pp. 6-7). Hospitals, medical colleges, and primary health care centers multiplied at an amazing rate. Infant and maternal morbidity and mortality were the first targets, and a significant downturn in these vital statistics was achieved.

India gained independence from the British in 1947. Celebrations were marred by the trauma of partition, and an unprecedented exchange of population, wanton killing, and destitution were enough to daunt faith in human decency and benumb ethical sensibilities. Violence and hatred between religious communities continue to plague India; sanctity of human life is sacrificed on the altar of religious supremacy.

In spite of the avowal of ideals of non-violence, what may be called communal madness has been endemic in India, the subcontinental evil in which victims are brutalized, no longer seen as human beings. From Buddha to Mahatma Gandhi, the apostles of non-violence have espoused the ideals of good and moral life, especially compassion. But even Gandhi could not stem the carnage at the time of partition. In the last few years Punjab, Gujarat, Assam, Delhi, and Bihar have been engulfed in the conflagration, unrelieved by explanations and unmitigated by appeals to religious tolerance and secularism.

There is also the daily violence that occurs in India, and such degradation as rape, blinding, flogging, and burning alive are not rare. These acts are rationalized as retaliation or retribution, a situation not unique to India. I am not suggesting that the ideal of non-violence is proclaimed dishonestly, but a confluence of events can convert ordinary men into frenzied killers. It may appear that I am straying from my principal theme, yet an understanding of violence not only sets a context of human sensibilities, but also pinpoints psychosocial realities that should be squarely in the domain medical ethics.

PROBLEMS OF MEDICAL ETHICS

Ethical dimensions of medicine in India stand in stark contrast to those in the West. Western ethicists are confronting novel situations of life-and-death problems created by modern technology. Issues related to public health, for example, infant mortality, malnutrition, health problems determined by race, as well as problems of family disorganization and violence created by poverty and discrimination, are receiving scant attention from medical ethicists. An Indian ethicist must confront the problems of poverty, health education, and preventable death and diseases. For example, in their critique of modern medicine, Nandy and Viswanathan [55] focus mainly on the colonial dwarfing of native traditions and point to the failures of a modern medicine which is oriented to curing rather than to preventing, a la Ivan Illich [34]. They join a chorus of health care professionals in India who are pushing for a dramatic shift in the delivery system. The goals established by the Bhore Committee have not been realized, nor has there been a serious attempt to implement them all [27]. Those who indict the current system fix the responsibility on the logic intrinsic to Western medicine. They allege that it is a science born out of dominance, first over nature, setting humans apart, and then disconnecting the patient from the physician, whose objective knowledge and skills must dominate the subjective experience of the patient. Western medicine tied itself to the coattails of colonialists and imperialists; and showed similar disdain for what was native. Cast in chauvinistic language, this critique, drawing from the sociology of knowledge, has its appeal in pointing to gross problems of inequity and neglect. What may be drowned in this rhetoric against modern medicine are the contributions of science in public health, especially epidemiology, that facilitate combating preventable morbidities. I, too, found in a recent consulting experience with the government of India that the medical establishment opts for bricks and mortar, expensive tertiary care facilities and medical colleges, and that this view obscures the cultural conditions that promote hierarchically determined elitist solutions.

DRUGS, DOCTORS, AND DEVILS

Another fallout from the tradition of imperialistic medicine is the pharmaceutical industry. The campaign in the third world against baby foods and breast milk substitutes amply demonstrated that the contentions were valid, and that a wholly inappropriate substitution of breast-feeding was being encouraged in the third world. In the West the women's movement had already eroded markets for these products; breast-feeding had been restored as a healthy feminine activity and in the best interest of the infant. Although the offending company agreed to alter its advertisement to indicate that mother's milk is best for an infant, and stopped distributing free samples to maternity wards, the issue continues to concern those in developing countries. An Indian company is now under a similar attack [73]. Transfer of health and nutrition technology is being hotly debated as well [30].

The pharmaceutical industry is truly an international operation, and several of its concerns have become bones of contention in India. On the one hand, some multinational companies are withholding the marketing of new drugs in India because of disputes over patent rights laws, and on the other, drugs posing high risks before approval or after rejection by agencies like the FDA in the U.S. are introduced in the third world. Stringent guidelines to protect human subjects in experimental investigations in the U.S. and other Western countries make human trials cumbersome, and so drug companies have resorted to such trials in the third world ([21]; [26]; [64]).

The local drug companies in India, some of them subsidiaries of multinationals, may be seen to reflect the confrontation between medical ethics and what is for the West early technology. Thousands of useless over-the-counter drugs have proliferated, glutting the market and diverting scant medical resources ([15]; [29]; [80]). Questionable prescription drugs, often available through the local pharmacist without a doctor's prescription, mirror a medical practice in which a physician prescribes for a patient after listening only to a messenger's (often a family member's) description of that patient's symptoms. The pharmacist becomes a kind of a doctor, which for the patient circumvents the need to go to a physician, and saves time and money.

Despite the oversight mechanisms which are now in place, there are numerous problems. The Lenten Commission, appointed by the government of Maharashstra to investigate a rash of deaths from the administration of a contaminated medication in a major teaching hospital in Bombay, concluded with a wholesale indictment of the regulatory system. The nexus between the drug industry, the bureaucrats, and the politicians, according to the commission report, made a mockery of regulation, since money changing hands made any irregularity acceptable [83]. In other reports physicians are not held blameless either, and are suspected of playing to the tune of drug companies ([6]; [14]). It is more the rule than the exception that a patient consulting a specialist will return with a cocktail of prescriptions, each medication targeted for each symptom, and some drugs of questionable value. Hormones, vitamins, elixirs (or tonics, as they are called in India), and tranquilizers are prescribed without restraints or appropriate indications ([21]; [26]). To add insult to injury, even drugs established by medical evidence to be harmful, and banned elsewhere, are not taken off the market [15]. On some occasions when I have reviewed the treatment of an Indian patient, I have been amazed at the number of filled prescriptions. It is a sorry spectacle of ethical failures at multiple levels, rooted in a native medical system of the patient and the physician that is symptom-oriented, each symptom deserving a drug, often without or with only a cursory examination, and served by a pharmaceutical industry, both multinational and fly-by-night local, that is aided and abetted by regulators, politicians, and physicians [14]. Drugs, doctors, and devils combine to unravel an already tattered system of medical care.

REPRODUCTIVE TECHNOLOGY

Absent modern technology, new reproductive technologies have not been at issue in modern India. Some Indian laboratories have reported in vitro fertilization with successful outcomes. Women are reported lining up to receive IVF. Everything in the law books and the medical texts suggests that all methods of enhancing conception are to be sought and tried. The stigma of infertility, especially for women, has always been severe, and all means are sought to secure progeny, particularly male children.

Niyoga, lawful cohabitation of a childless wife with her brother-in-law, and failing that, with a virtuous brahmin, was well established in ancient times. There are many communities in modern India where the practice is commonplace ([40], pp. 536-556). Thus, semen donation has been practiced, albeit through more traditional means, and rather than being an impediment is a moral imperative (*Mahabharata* 1.7.113.15-25 [49], p. 254). It should be possible to construct moral paradigms to make the case for artificial reproductive technology with the aid of a mythology which has glamorized a variety of instances of ovum donation, embryo transplant, and rules of adoption (mainly for the purpose of inheritance) ([20], pp. 64-68). In India there always will be a preference for known donors, thus ensuring correct blood lines, over anonymous transactions. The unpopularity of adoption testifies to the aversion to accepting babies of unknown parentage.

WOMEN, ABORTIONS, AND AMNIOCENTESIS

Over the last several years, amniocentesis has been the most publicly debated crisis of medical ethics in India. Daily newspapers, national as well as the regional, have focused attention on how a modern technology can awaken and make explicit conflicts inherent in a culture. Sunday supplements and popular periodicals have diverted space to an explanation of the procedure [32], and have encouraged the airing of tensions between traditional and modern values [69]. A deeper appreciation of the problem requires an understanding of the place of women in Hindu society.

It is difficult to determine exactly and at what point in the development of the Hindu tradition the lawgivers attached greater value to the lives of men than to those of women. It is certain that by the time the *Laws of Manu*, the best known and almost universally recognized Hindu law text, was written, women were regarded with apprehension. A more restrictive code of conduct was formulated, and it was declared that women should be not allowed independence; as young girls they should be under the command of their fathers; as wives, of their husbands; and in old age, of their sons (*Manu* 5.148 [50] p. 195; 9.3, p. 328). A father must arrange the marriage of his daughter before menarche or invite the sin of feticide with every passed menstrual period ([38], II, pp. 438-446); every missed opportunity for conception is a crime against society. Explicit in the injunction was the recognition that, with the coming of puberty, the fires of sexuality will lead a young girl astray.

There is evidence that the earliest Indian civilization was matriarchal, and that the worship of mother goddesses was common. Unlike other ancient societies, the coming of Indo-Europeans did not completely transform the Indian civilization into a patriarchal organization. We have already noted that the dispersal of population from India's heartland was slow; moreover, the hegemony of the brahmins, the purveyors of tradition, was variable. Mother goddesses of earlier worship traditions were incorporated into the Hindu pantheon. Even in the iconic and mythological representation of the female goddesses a bivalence persisted, a nurturant and protecting mother versus a destructive and frightening goddess. In modern India the tradition is best exemplified in the worship of Durga and Kali in Bengal. In the Southern state of Kerala, and in some other parts of India to lesser degree, a matriarchal residence and property transmission (from mothers to their daughters) has been the norm. One may speculate that the tensions between the two original cultures have persisted in these attitudes ([42], pp. 42-108). It would appear that the anxiety over the sexuality of females led to social restrictions on their movement. It is also likely that the nomadic life of the Indo-Europeans was transformed +wto settled agricultural pursuits in the Indo-Gangetic plains, bringing a consciousness of property rights. Inheritance laws created a need for fixing paternity. The law books are unremitting in their condemnation of miscegenation and, generally, the mixing of blood lines between castes, because the progeny so produced sink down the hierarchical order and are consigned to lowly tasks (*Manu* 3. 12-13 [50], p. 77).

There was one important exception: the stage of householdership was celebrated as a time when a man discharged his ritual obligations to gods, guides, and fathers. For the attainment of these goals, a man needed a wife to give him sons. Motherhood, especially of a male child, enhanced a woman's self-esteem and social status; her contribution to generational continuity empowered her in family politics. As a man needed to fulfill his obligation to his father, so did he need a son who, through the performance of a sacred rite called *shraddha*, rendered him whole after his death and assured a passage to the land of the fathers. This special aspect of the father-son relationship involves building a symbolic body for the dead father with rice balls, to which are added smell, sight (color and form), clothing, and finally breath; the son reconstitutes the father by giving his own body to his dead father [41]. The son enhances the father by pouring himself into the bodiless dead father, in return for the father's having once poured himself into the act of conception. The rite is a prime obligation for the son, woven into Hindu eschatology, and considered by many as the source of the yearning for a son.

The physical and psychological well-being of women has long been in jeopardy. The notorious medieval rite of *Sati*, the self-immolation of widows on the funeral pyres of their husbands, was banned by the British in 1829, with help from Raja Ram Mohan Roy. Roy, a Hindu reformist, appealed to values outside the Hindu tradition to protest the practice of widow burning, which was particularly rife in 18th-century Bengal ([38], II, pp. 624-636). A recent episode of *Sati* in 1987 created a national stir and led to new legislation that makes it a crime to abet or exalt the act of *Sati*. This isolated instance may have been a cultural assertion of Hinduism in response to an earlier piece of legislation that allowed Muslim religious laws regarding divorce and maintenance of divorced women to prevail over the civil law. In essence it may be a Hindu way of saying, "If the Muslims in India are allowed to treat their women according to their laws, we shall treat ours according to our religious law".

Neglect and deliberate undernutrition of female babies and female infanticide have also been known in India. Barbara Miller [54] has extensively reviewed the practice in North India. More recently, Amartya Sen analyzed the adverse ratio of women to men in certain parts of the world, particularly India (0.94 to 1), and suggested that an interplay of economic and cultural conditions are responsible for that ratio ([75], pp. 61-66).

In this context, let us review the issue of contraception and family planning. With modern medicine, e.g., antibiotics, blood transfusion, aseptic surgery, immunization, etc., and an explosion in health services after independence in 1947, India's population also began to explode. In the face of the consequent declining death rate and an unabated birth rate, India embarked on a national family planning campaign. The debate that ensued against family planning was not about ethical constraints against contraception. It was phrased in sectarian and caste terms: the lower castes and religious minorities would grow at a relatively faster rate and adversely affect the interests of the Hindus, especially the upper castes, who because of their educational levels would more likely restrict the size of their families (a claim which has not been born out by subsequent events).

In Hindu tradition and ancient Hindu medicine, contraception had not been contemplated. But means sought to enhance conception were plentiful. From appropriate matching of couples, recommendations about timing of intercourse and seed and conception-enhancing feeds, special procedures and medications for sexual rejuvenation called *vajikarana* (lit.,"making like a horse") (*Car.* 6.2.3 [13], pp. 35-52), and a permissive attitude about sexual congress (especially for men of higher castes with women from lower, and for men about town ([37]), it is only a small jump to contraception. The 16th-century medical text *Bhavaprakasha* advises a few oral preparations to prevent conception (*BP* 2.71 [11], Secs. 32-33, p. 1525).

In recent years male and female sterilizations have been performed in the millions, with monetary rewards for those who opt for these methods. Recently, a comprehensive insurance scheme has been proposed by the Indian Medical Association, covering sterilization and its complications, to encourage family planning [65]. Distribution of free condoms is ubiquitous, and every known method has been in vogue, except perhaps the pill [7]. Indians are psychologically more sensitive to ingested than to surgical interventions. On the other hand, it is generally believed that during a period known as the "emergency" in the seventies, the then government of Indira Gandhi so heightened the tempo of sterilizations, especially of males, that the issue was implicated in a major election defeat ([1], pp. 188-194; [12], pp. 126-127). Thereafter, politicians avoided talking about family planning, and even a central government ministry was renamed, from the Ministry of Family Planning to the Ministry of Family Welfare. Apart from the excesses of forced or inappropriate vasectomies, an existing but, until then, marginal belief gained strong currency that such procedures produced impotence. Family planning programs have not recovered yet.

Ayurvedic medical texts dwell explicitly on the subject of conception. From the earliest *CarakaSamhita* to the latest *Bhavaprakash*, there is an agreement that life begins at conception. The older texts regard the male and the female generative substances as containing four of the five elemental substances (fire, water, wind and earth), the fifth (*akasha*, ether or space) entering the process of fertilization as self or consciousness (*Car.* 4.2 3-4 [13], p. 412; *Su.* 3.3 2-3 [81], p. 436; *Ahr.* 2.1-2 [3], p. 210). Later texts speak of all the five elements as part of the human seed, thereby making *akasha* more material and less ethereal. Early medical texts debate the factors that go into the origin of life, and conclude that parents, nutrition, timing, and the self are independent variables. No factor alone can determine conception (*Car.* 4.3 [13], pp. 418-427). Later texts attempt to remove the mystery of the emergence of the self at the moment of conception by an analogy: just as rays of the sun combining with (passing through) a lens reveal a hitherto unknown power, that of fire, so the self is manifested from the combination of male and female germinal substances (*Ahr.* 2.1.3 [3], p. 210). Signs of pregnancy are recognized very soon after conception, and the mother and the embryo begin to influence each other from the very beginning.

But is this human life? Hindus would argue that this is a Western question. Hindus have built a cosmology in which the Judeo-Christian distinction between humans and other creation simply does not exist. The self, often the equivalent of what is considered soul in the Western tradition, is a part of the creator and is the essence of all creation. Humans distinguish themselves from other created beings as adults by their moral conduct. On the other hand, Ayurvedic embryology delineates different stages of development with distinct implications for corresponding maternal awareness and attachment. The heart is considered the seat of the soul. It is given a special status in development over the growth of other body parts, and is equated with the moment of quickening, which marks the coming into being of a formed self from an earlier amorphous fetal mass (*Car.* 4.4.10-15 [13], pp. 430-431; *Su.* 3.3.17 [81], p. 441). Ayurvedic texts describe development in the eighth lunar month (28 to 32 weeks) as critical for both mother and fetus, with the life principle (*ojas*) moving to and fro between the two, making the viability of the fetus uncertain, and endangering maternal life (*Car.* 4.4.24 [13], p. 433; *Su.* 3.3.14 [81], p. 440).

Ancient Indian law considered abortion a sin, and medical texts warned against medical or surgical interventions that could lead to loss of pregnancy (*Car.* 4.8.19-32 [13], pp. 432-435). The religious texts considered abortion as a high crime, equating it with the killing of a brahmin ([38] II, p. 148n; III, p. 411n, p. 612n). Although the evidence for this equation appears to arise from a linguistic equivalence between the word for a brahmin and for a fetus whose sex was not known, women resorting to abortions were despised. It is apparent from the consideration of threatened abortion in the medical texts that attempted abortions were known to the physicians at that time (*Su.* 2.8.2 [81], p. 377). In cases of obstructed delivery, the *SushrutaSamhita* advises the physician not only to be very careful, but also seek the permission of the royal court, for loss of life or lives was an imminent danger. Surgical removal of the fetus is elaborately described, with the admonition not to intervene until the fetus is dead. In the end, however, the text favors saving the mother's life (*Su.*4.15 [81], pp. 645-650; *Ahr.* 2.2.29-39 [3], pp. 224-225). With the passage of time, medical means came to be prescribed for the termination of early pregnancy. A 17th-century physician advocates a preparation for vaginal use to precipitate labor ([16], p. 45).

In modern India the foregoing is history. In 1972 a comprehensive Medical Termination of Pregnancy Act was passed, and abortion has been available on demand since that time. The law places some restrictions on termination between the 14th and 20th week, but failure of traditional contraceptive measures is one of the indications ([16], pp. 82-108). When the law was passed, opinion makers, especially in the popular media, overwhelmingly supported the measure, not only as a method of population control, but also to end illegal abortion mills. The meager opposition came mainly from the Catholic press ([16], pp. 146-160). Today newspapers, all of which have at least a weekly section on health, educate people on issues like safe abortions [74]. RU 486 will soon be offered in clinical trials [79]. In this regard, Julius Lipner does not seem to capture the Indian ethos in his recent paper on this issue [48]. Although he can call to his aid significant ancient authorities condemning abortion, he ignores the contextual nature of Hindu ethics, its reliance on conduct and local tradition, and the advocacy that preceded and followed the enactment of the Medical Termination of Pregnancy Act. Most importantly, he fails to see the distinction in ancient medical texts between the moral status of the unborn and that of a woman resorting to abortion. Absolutist interpretation of scriptural injunctions does not have the force in modern India which it does elsewhere. Indian independence and the subsequent secular democratic tradition required abrogation of many religious strictures, untouchability and women's rights foremost among them.

Legally available abortions in India are there to stay. Their success as a family planning method is neither impressive nor celebrated. Recourse to medical abortion is a boon to those who would otherwise seek an illegal one to avert the stigma of pregnancy, and such attitudes about abortion are changing. One can hardly lose sight of the fact that the texts which inveigh against abortions are notorious for their misogyny (e.g., *Manu*. 10.1-67 [50], pp. 401-417). They interpret abortion as permission to women for licentiousness. One does not have to stretch one's imagination to see that the lawmakers banned abortion to bring down the full weight of social scorn on women pregnant out of wedlock.

All of this brings us to amniocentesis. Modern technology for fetal diagnosis of congenital conditions also permits determination of sex. In India the epidemic of female feticide has drawn a loud protest, especially from women activists [69]. Newspaper accounts have estimated that 18,000 abortions between 1978 and 1983, and 78,000 between 1983 and 1988 in Bombay alone, were linked to gender preferences [45]. Advocates who find the practice permissible argue that if abortions are legal, people should also have the option of choosing the sex of their child ([22]; [33]; [36]; [43]). The tradition overwhelmingly prefers, even mandates, having sons, and a female child is burdensome because of the dowry system, among other reasons. Advocates point to dowry deaths, when young brides are killed by the in-laws, either because they did not bring enough of a dowry, or because killing them permits remarriage and a new dowry ([43]; [78]). Those who oppose abortion for the purpose of sex selection argue against it as discriminatory and barbaric, as reducing the humanity of the woman, and as further aggravating the already adverse demographic patterns of male-female ratios ([4]; [39]; [45]; [47]; [82]). The traditional belief that a woman determines the sex of a child has also been challenged with a poster campaign to educate masses about male and female chromosomes in determination of a gender [64]. A group of gynecologists has justified the use of ultrasound for fetal sex determination [60], and one physician who performs the test argues that it is better for an unwanted girl not to be born [57]. Others have argued that women are not commodities ([71]; [78]), and that the misuse of technology undermines medical professional ethics. Some have proposed a boycott of "sex-test docs" [25], and others have organized public demonstrations ([24]; [66]). In response to intense debate and agitation, the government of the state of Maharashtra passed a law in April 1988 placing severe restrictions on amniocentesis, outlawing the procedure for the purpose of sex-selection (with the physician punishable by a fine and/or a prison term), and permitting it only when well established suspicions of congenital malformations exist ([59]; [62]). The movement to curb the procedure nationwide has not succeeded, although proposals have been made ([61]; [67]), and the difficulty here remains enforcement [76]. Like most other secular laws that are easily and flagrantly violated, it is unlikely that an entrenched tradition in which sons are preferred will disappear, so long as abortions are legal; it will simply go underground, like the abortion mills of an earlier era. There is nothing in Hindu tradition or culture that can be invoked to save female fetuses, and moral pressure on parents or doctors on the basis of sanctity of human life is of little avail in dissuading them when the couple already has one or two female children. Once reproductive technologies in India make it possible to assume and enhance conception of a male baby, rivaling the Hindu rite of *pumsavana* ([38], II, pp. 218-220), which advises procedures and preparations for the development of the fetus of a desired gender in keeping with the Ayurvedic medical texts (*Car.* 4.8.9.12 [13], pp. 463-465; *Su.* 3.2, 14-15 [81], pp. 430-431), perhaps amniocentesis will lose favor.

DEATH, DYING AND EUTHANASIA

Vedic Indians celebrated life, and not until the time of Upanishads had the melancholy of death set in. Vedic prayers ask for long life, and call upon the Gods to keep death and old age away. The Upanishadic literature may be regarded as a response to the anxiety aroused by the finality of death and narcissistic dissolution. The soul or the self (*atman*) became the real and transcendental entity, and death was defined as only that of the body. To wish the moment of one's death became a spiritual achievement, granted to some as a boon and achieved by some others through yogic practices. An understanding of death in the scriptural literature of India has been comprehensively presented by Katherine Young ([84] pp. 71-130).

In the medical texts, the ancient physicians argue that being alive is all important (*Car.* 1.11.3-4 [13], p. 69). They advise the physician not to attend to incurable diseases (*Car.* 1.1.62 [13], p. 8), and warn the physician that in "treating an incurable disease [the physician] certainly suffers from the loss of wealth, learning and reputation and from censure and unpopularity" (*Car.* 1.10.8 [13], p. 67).

It is not difficult to visualize how the ancients would have tackled the problems of dying. Their intent clearly appears to be to permit those without hope to die. *Caraka* advises that active treatment should be withheld from a dying person, and feeding should be continued only for the comfort of the relatives (*Car.* 5.9.14-17, [13], p. 516). More difficult to visualize are acts of active assistance in the process of dying. Around the abortion controversy, Lipner has raised the issue of the principle of non-violence, *Ahimsa*, so central to Hindu (and Jain and Buddhist) thought, and has invoked this principle against abortion in India ([48], p. 59). There cannot be a better modern interpreter of *Ahimsa* than Mahatma Gandhi, and it is pertinent to review what he has said on the issue of euthanasia. Bhikhu C. Parekh, who has presented a penetrating analysis of Gandhi's intellectual life, discusses an episode in Gandhi's life, in which a calf which had no hope of surviving was put to sleep. He cites Gandhi as rejecting the view "that killing was never justified and that all killing was violence" ([70], pp. 120-121). Gandhi said that there is violence when the intention is to give pain; otherwise, it is only an act of killing ([70], p. 121). For Gandhi, to have freed the calf from suffering was an act of love. Gandhi argued that his wealthy Hindu and Jain critics committed far greater violence by paying their laborers sub-standard wages, because there is "far greater violence in the slow torture of men...and wanton humiliation and oppression of the weak and the killing of self respect...than in mere benevolent taking of life" ([70] p. 121). Gandhi was confronted by his critics with the problem of dying patients, and his response is summarized by Parekh in the following way:

1. The disease from which the patient is suffering should be incurable.
2. All concerned have despaired of the life of the patient.
3. The case should be beyond all help or service.
4. It should be impossible for the patient in question to express his or her wish.
5. So long as even one of these conditions remains unfulfilled, the taking of life from the point of view of Ahimsa cannot be justified ([70], p. 122).

Although Gandhi believed that he had arrived at his position independently, he was building here on the position advanced by ancient medical authorities.

OTHER ISSUES

There are a few other issues that have generated debates about medical ethics. The first of them is medical ethics itself. The Indian Medical Council formulated a code of ethics in 1972. The current debate focuses mainly on the commercialization of the practice of medicine. The participation of physicians in drug company advertisements and in the endorsement of products draws fire from the popular press and public opinion ([6]; [28]). Social commentators, including physicians dedicated to a public health care delivery, express their disappointment with the lack of moral commitment by physicians to the service of the poor and the rural masses ([2]; [8]). Expectations that physicians will practice with the highest medical standards, putting the interests of patients first, and devoting part of their time to *pro bono* service, are, however, cries in the wilderness. Among the codes of ethics is also an annotation that implores the physician to respect and honor his or her elders, thus revealing the cultural dimension of a hierarchical tradition where authority requires deference even in matters scientific [53].

A second debate centers on organ donations for transplants. A few years ago I suggested that the matter of selecting a donor for an Indian patient requires further investigation, especially regarding issues of social and psychological matching. I believe the question has been answered; the way "organ bazaars" [19] are procuring organs for sale to the highest bidder removes any doubt about what is important [18]: to the recipient, a healthy organ that can prolong life; to the donor, monetary compensation; and to the hospital and the physician, money [63]. The scandal calls for self-regulation; but if the future of a whole family is changed by the sale of an organ fetching more money than what the family might expect to earn in a lifetime, it is difficult to take a moral or legal stance against such a transaction. On the other hand, there are potentials for abuse in substandard surgical interventions, inappropriate screening, and even coercion, which the medical profession can best police. The tradition of donation otherwise is rare [27].

Technology is simply not available to care for severely premature babies. In view of a high infant mortality rate and the concentration of medical resources in urban areas, where less than 25 percent of the population live, it is unlikely that neonatology will save many babies, although feticide for reduction of multiple pregnancies may present problems [68]. Medical research is poorly funded. The specter of fetal tissue for research is but a distant possibility, unless, of course, like organ donation, an international trade arises. In tune with animal rights, the Government of India has banned the export even of monkeys for research.

Confidentiality deserves mention only because the general practice of medicine, both modern and traditional, is conducted in storefront offices, where the physician questions and superficially examines the patient in full view of other patients. Specialists, referred to as consultants in India, have private examining rooms, but this is more a matter of status than in the interest of confidentiality. The downside of the situation is that patients are wary of disclosing intimate details. They often avoid consulting psychiatrists, because they are afraid that their secrets will become public.

Finally, in the public ethical debates conducted in newspapers, controversial views are not usually presented in the context of modern medicine vis-a-vis traditional Hindu values. More often than not, Hindu values are attacked, and partisanship is expressed for democratic rights to secular dignity and equality of access. Newspaper columnists keep an eye on developments in the West, especially the United States and the United Kingdom, and often report on landmark legal cases involving medical ethics ([5]; [44]; [46]), not unlike Indian jurists who to refer to legal decisions in the West to formulate arguments and judgments in Indian courts.

The task confronting a medical ethicist in India is to extrapolate from both ancient and modern consciousness that which might guide one through ethical dilemmas. The Hindu tradition permits a reinterpretation of ancient wisdom in the context of changed realities. After all, ancient religious and medical values were developed in times when average life spans were short, death rates, especially those of infants and mothers in labor, were high, and external threats led to higher birth rates in order to increase the size of tribe or group. What the ancients might have done in a world where technology can successfully alter processes of conception, birth, and death can be addressed only from their ontological and epistemological principles, as well as their constructs of the self and the body, of human interaction with the environment, and of pain and disease.

University of Illinois Medical Center
Chicago, Illinois, U.S.A.

BIBLIOGRAPHY

1. Ali, T.: 1985, *The Nehrus and the Gandhis: An Indian Dynasty*, Pan Books, London.
2. Antia, N.H.: 1989, "Poor Health Services: Who is to Blame?", *Economic and Political Weekly* (September 30), India, 2214-2215.
3. *Ashtanqhridaya*: 1983, (text with Gujarati translation), trans. V.D. Munshi, *Sastu Sahitya Vardhak Karyalaya*, Ahmedabad, India.
4. Baig, T.A.: 1988, "Gender Genocide", *Miscellany* (May 8), India.
5. Bakshi, P.M.: 1989, "Test-Tube Babies and Legitimacy", *Hindu* (January 25), New Delhi.
6. Bal, A.: 1988, "Medical Ethics" (letter to the editor), *Indian Express* (April 14), Bombay.
7. Balasubramaniam, V.: 1988, "Pushing the Pill", *Indian Express* (October 9), Bombay.
8. Banerji, D.: 1989, "Crisis in the Medical Profession", *Economic and Political Weekly* (March 20), India, 1091-92.
9. Banerji, D.: 1976, *Formulating An Alternative Rural Health Care System For India: Issues and Perspectives*, Centre for Social Medicine and Community Health, Jawaharlal Nehru University, New Delhi.
10. Basham, A.L.: 1959, *The Wonder that was India*, Grove Press, New York.
11. *BhavaDrakash*: 1981, trans. (in Gujarati) G.M. Shastri, *Sastu Sahitya Vardhak Karyalaya*, Ahmedabad, India.
12. Brata, S.: 1985, *India: Labyrinths in the Lotus Land*, William Morrow and Company, Inc., New York.
13. *CarakaSamhita*: 1981-83 (2 vols.), trans. P. Sharma, Chaukhambha Orientalia, Varanasi, India.
14. Chakravarty, S.: 1988, "The Unholy Trinity", *Sunday Observer* (April 3), India.
15. Chandra, S.: 1988, "Chemists Flourish on Banned Drugs", *Times of India* (April 27), Bombay.
16. Chandrashekhar, S.: 1974, *Abortion in a Crowded World: The Problem of Abortion with Special Reference to India*, Allen and Unwin, London.
17. Chattopadhyaya, D.: 1977, *Science and Society in Ancient India*, Research India Publication, Calcutta.
18. Colabawala, B.N.: 1989, "Kidneys for a Price", *Indian Post* (February 23), Bombay.
19. Cover Story: 1990, "The Organs Bazaar", *India Today* (July 31), India.
20. Desai, P.N.: 1989, *Health and Medicine in the Hindu Tradition*, Crossroads, New York.
21. Devarajan, P.: 1990, "Drug Policy: A Case of Mixed-Up Priorities", *Financial Express* (March 6), New Delhi.
22. D'Souza, J.B.: 1988, "Should There Be a Choice?", *Times of India* (January 10), Bombay.

23. Dumont, L.: 1980, *Homo Hierarchicus*, University of Chicago Press, Chicago.
24. Dutt, N.: 1988, "Women Groups Demand Ban on Sex Test", *Indian Express* (December 27), Delhi.
25. Gandhi, N.: 1988, "Boycott of Sex-Test Docs Urged", *The Hindustan Times* (February 28), Delhi.
26. Ghosh, A.: 1988, "Drug Policy: An Outline for the Future", *Economic and Political Weekly* (April 16), India, 766-67.
27. Ghosh, S.: 1989, "Body Donation", *Times of India* (February 25), Bombay.
28. Ghoshal, S.: 1988, "Outdated Concepts of Medical Ethics", *Indian Express* (April 16), Delhi.
29. Gomes, J.: 1988, "Blinding Drug on Sale", *Midday* (May 24), India.
30. Gopalan, G.: 1990, "Health/Nutrition Technology Transfer", *Economic and Political Weekly* (January 27), India, 223.
31. Gould, H.: 1965, "Modern Medicine and Folk Cognition", *Human Organization* 24, 201-8.
32. Grover, A.: 1986, "Amniocentesis or Female Feticide?", *Lawyers* (April), India.
33. Hansoti, R.C.: 1989, "We Should Produce Only Children of the Desired Sex" (letter to the editor), *Post* (March 14), India.
34. Illich, I.: 1976, *Medical Nemesis: The Exploration of Health*, Pantheon, New York.
35. *India*: 1989, Research and Reference Division, Ministry of Information and Broadcasting, Government of India, New Delhi.
36. Jain, K.: 1988, "Why Blame Amniocentesis?" (letter to the editor), *Indian Express* (March 23), India.
37. *Kamasutra of Vatsayana*: 1964, trans. F. Burton, E.P. Dutton, New York.
38. Kane, P.V.: 1968-77 (1930-62), *The History of Dharmashastra*, Bhandarkar Oriental Research Institute, Poona.
39. Kapur, R.: 1988, "Sex Test Ban" (letter to the editor), *The Times of India* (December 20), Delhi.
40. Karve, I.: 1962, "The India Social Organization: An Anthropological Study", in S.K. De *et al.* (eds.), *The Cultural Heritage of India*, Vol. 2, Ramakrisna Mission, Calcutta.
41. Knipe, D.: 1977, *SapindaKarana*: "The Hindu Rite of Entry into Heaven", in F.E. Reynolds and E.H. Waugh (eds.), *Religious Encounters with Death: Insights from History and Anthropology of Religions*, University of Pennsylvania Press, University Park, pp. 111-124.
42. Kosambi, D.D.: 1962, *Myth and Reality*, Popular Prakashan, Bombay.
43. Kumar, D.: 1988, "Ban on Sex Test Clinics Unwise", *The Times of India* (December 9), India.
44. Kusum, D.: 1989, "Who Has a Say?", *Hindustan Times* (July 21), India.
45. Kusum, D.: 1988, "Are Indian Women Becoming an Endangered

Species?", *Times of India* (December 27), New Delhi.
46. Kusum, D.: 1986, "Of Fetal Rights", *Hindustan Times* (July 27), India.
47. Letters to the Editor: 1989, "D. Hansoti Receives Flak for Advocating Female Feticide", *Post* (March 23), India.
48. Lipner, J.J.: 1989, "The Classical Hindu View of Abortion and the Moral Status of the Unborn", in *Hindu Ethics: Purity, Abortion and Euthanasia*, ed. H. Coward *et al.*, State University of New York Press, Albany.
49. *Mahabharata*: 1973, trans. J.A.B. Van Buitenen, University of Chicago Press, Chicago.
50. *Manusmriti (The Laws of Manu)*: 1969 (1886), trans. G. Buhler, Dover Publications, New York.
51. Marriott, M.: 1969, "Little Communities in an Indigenous Civilization", in *Village India: Studies in the little Community*, ed. M. Marriott, University of Chicago Press, Chicago.
52. Marriott, M., and R. Inden: *Encyclopedia Britannica*, 15th ed., s.v. "Caste System".
53. Medical Council of India: June 1980, "Code of Medical Ethics", supplement to *C.M.A.I. Journal*, India.
54. Miller, B.: 1981, *The Endangered Species: Neglect of Female Children in Rural North India*, Cornell University Press, Ithaca.
55. Nandy, A. and S. Viswanathan: 1990, "Modern Medicine and its Nonmodern Critics: A Study in Discourse", in F.A. Marglin and S. Marglin (eds.), *Dominating Knowledge,* Clarendon Press, Oxford.
56. Nehru, J.M.: 1981 (1946), *The Discovery of India*, Jawaharlal Nehru Memorial Fund, Delhi.
57. News Report: 1988, "Better for Unwanted Girl Not to be Born", *Indian Express* (January 29), India.
58. News Report: 1988, "Prevarication on Harmful Drugs", *Economic and Political Weekly* (April 16), India, 756-57.
59. News Report: 1988, "Bill Banning Pre-Natal Sex Test Passed", *Indian Express* (April 17), India.
60. News Report: 1988, "Ultrasound Use Finding Fetal Sex Justified", *Times of India* (April 18), Bombay.
61. News Report: 1988, "Central Ban on Sex Test Soon", *Times of India* (June 6), Delhi.
62. News Report: 1988, "Bill Introduced", *India Express* (September 14), Bombay.
63. News Report: 1988, "Racket in Kidney Transplants", *Statesman* (November 7), India.
64. News Report: 1988, "Poster Campaign on Sex of Child", *Indian Express* (December 27), Bombay.
65. News Report: 1989, "Insurance Scheme for Sterilization Mooted", *Indian Express* (April 18), India.
66. News Report: 1990, "Protest against Sex Discrimination Test", *Indian*

Express (March 8), Bombay.
67. News Report: 1990, "Government Silent on Sex Determination Test", *Indian Express* (May 4), Bombay.
68. News Report: 1990, "Selective Abortion May Spark Debate", *Times of India* (May 30), Bombay.
69. Nilekani, R.: 1988, "Giving Birth to Trouble", *Sunday* (June 12), Delhi.
70. Parekh, B.C.: 1989, *Colonialism, Tradition and Reform: An Analysis of Gandhi's Political Discourse*, Sage Publications, New Delhi.
71. Poonacha, V., and V. Lalwani: 1989, "Women are Not Commodities", (letter to the editor), *Sunday Observer* (March 26), India.
72. Pranvir, A.K.: 1990, "Health Care: Profitable Business", *Hindustan Times* (February 5), India.
73. Rangan, K.: 1988, "Consumer Body Plans Boycott of Amul", *Indian Post* (October 8), India.
74. Rao, B.K.: 1990, "Risk Factors in Abortion", *The Hindu* (May 20), New Delhi.
75. Sen, A.: 1990, "More Than 100 Million Women Are Missing", *New York Review of Books* 37:20 (December 20), 61-66.
76. Setalvad, T.: 1988, "Just Another Paper Law", *Indian Express* (April 18), Bombay.
77. Sharma, P.H.: 1968, *History of Ayurveda*, trans. G.M. Shastri (*Ayurvedno Itihas* in Gujarati, from Sanskrit and Hindi), *Sastu Sahitya Vardhak Karyalaya*, Ahmedabad, India.
78. Sharma, K.: 1988, "The Problem of Girls", *Indian Express Magazine* (December 8), India.
79. Spotlight: 1988, "An 'Abortion Pill' for Indian Women", Asian News International", *Spotlight*, Chicago.
80. Srinivas, A.: 1990, "Drug Policy: Consumer's Loss, Industry's Gain", *Financial Express* (May 19), India.
81. *SushrutaSamhita*: 1973 (text with Gujarati translation), trans. K.G. Shastri, *Sastu Sahitya Vardhak Karyalaya*, 2 vols. Ahmedabad, India.
82. Taneja, N.: 1988, "Female Feticide No Democratic Right", *The Times of India* (December 26), Bombay.
83. Torcato, R.: 1988, "*Dramatis Personae* in JJ Affair", *Indian Express* (March 30), Bombay.
84. Young, K.: 1989, "Euthanasia: Traditional Hindu Views and the Contemporary Debate", in H. Coward *et al.* (eds.), *Hindu Ethics: Purity, Abortion and Euthanasia*, State University of New York Press, Albany.

MASAO FUJII

BUDDHISM AND BIOETHICS

INTRODUCTION

Buddhist ethics consist of sets of guidelines that regulate daily life towards attaining the goal of enlightenment. In this sense, it is antithetical to analytical bioethics. In this report, I first trace these general principles, and only thereafter talk about the application of these principles to practical bioethical problems as they arise in the country where Buddhist ethics has the greatest effect upon the practice of modern medicine, Japan.

GENERAL PRINCIPLES OF BUDDHIST ETHICS

Buddhist ethics refers to a set of acts regulating daily life that help humanity towards the ultimate goal of transcendental enlightenment. According to the Buddha's teaching, a Buddhist rejects worldly values, tries to gain insights into reality, and finds true happiness in life.

The Buddha himself did not say anything about metaphysical questions concerning life, such as, "Does the human soul survive or die with the death of the body?". Rather, he kept quiet, because it was futile to try to answer.

He also used parables. The following is most popular and frequently quoted one, known as the Parable of the Poisoned Arrow. Suppose a man were pierced by a poisoned arrow, and his relatives and friends called a medical doctor to pull the arrow out and treat the wound. But the wounded man stopped him from acting, saying, "Wait a little. Don't pull it out until the following information is revealed. Was it a man or a woman who shot the arrow? Where was it coming from? Was it a big bow, or a small bow? Was it made of wood or bamboo? What was the bow-string made of? Was it made of fiber or of gut? Was the arrow made of rattan or of reed? What features did it have? If we wait for all the information, what will happen?" There is no doubt that before all the information could be obtained, the poison would take effect and the man would die. He urgently needs to have the arrow removed and its poison kept from spreading ([1], pp. 426-432).

The Buddha declared that one should gain insight into the reality of human misery, understand its causes, and search for a way permanently to eradicate it. This is known as the Middle Way (Pali, *majjhima patipada*). The Middle Way, as the basic principle of the Buddhist life, avoids the two extremes of hedonism and asceticism.

The Buddha presents the Middle Way in his first sermon at Banaras, known as the Discourse on the Turning of the Wheel of Dharma (Pali, *Dhammacak-*

kapavattana-sutta) ([3], Vol. I, pp. 10-14). The Middle Way consists of understanding the Fourfold Holy Truths and the Noble Eightfold Path.

The Fourfold Holy Truths may be expressed as follows:

1. Existence is suffering.
2. Suffering is caused by selfish craving.
3. Selfish craving can be overcome.
4. It can be overcome by following the Noble Eightfold Path (Pali, *ariya-atthangika-magga*). The steps to enlightenment in the Noble Eightfold Path are:

1. right understanding (Pali, *samma-ditthi*),
2. right thought (Pali, *samma-sankappa*),
3. right speech (Pali, *samma-vaca*),
4. right acts (Pali, *samma-kammanta*),
5. right livelihood (Pali, *samma-ajiva*),
6. right effort (Pali, *samma-vayama*),
7. right mindfulness (Pali, *samma-sati*),
8. right concentration (Pali, *samma-samadhi*).

This eight-fold path aims at promoting and perfecting three essential stages of Buddhist training and discipline, namely: (1) ethical acts (Pali, *sila*; Skt., *sila*), (2) concentration of mind (Pali, *samadhi*), and (3) wisdom (Pali, *panna*: Skt., *pnajna*) ([4], p. 301).

These three stages are considered to embrace all the aspects of Buddhist doctrine and practice. *Sila* includes three components of the Noble Eightfold Path, namely right speech, right acts, and right livelihood. *Sila* is therefore an indispensable basis for the moral and spiritual attainments of both the laity and monks of the Buddhist Order. In other words, no spiritual development is possible without these moral stages. Basic ethical conduct rests on the five precepts (*panca silai*): not to kill, not to steal, not to commit adultery, not to tell a lie, and not to drink intoxicants. These precepts are not commandments, but rather guidelines set forth by the Buddha, to be followed in that spirit. The verses of admonishment of the seven Buddhas, including Sakyamuni, who appeared in this world, reads: "Not to do any evil, to do all the good acts, and to purify one's mind; this is the teaching of all the Buddhas" ([2], p. 27). The first and the second verse are often explained as the expression of general morality, and the third verse as the expression of one's religiosity. This verse, together with the precept of respect of life, takes Buddhist ethics beyond the consciousness of good and evil.

Theravada and Mahayana Ethics

After the death of the Buddha, several Buddhist Council Meetings were held in order to compile the Buddhist teachings. At the Second Buddhist Council Meeting, held one hundred years later at Vaisali, the Buddhist Order broke into two factions according to the differences in interpretation of the Buddha's original teachings. The school of conservative orthodoxy or Theravada (*thera*, Pali for "elders"), such as the Sarvastivada, was a group of conservative senior monks who advocated a strict adherence to the precepts, and the Mahasamghika was a group of more progressive monks whose ideas were to develop later into Mahayana.

The differences between the two were small at first, but soon grew, and at the Third Great Council Meeting, held at Pataliputra under the reign of King Asoka (268-232 B.C.), the Sarvastivada established the orthodox doctrine and expelled the progressive new group's advocates as heretics. At the Fourth Great Council Meeting, under the patronage of Kushan King Kanishka (*c.* 150 A.D.), the divergence between the Larger and Smaller Vehicles, the Mahayana and Hinayana (which includes Theravada), was formally established.

These two branches of Buddhism have evolved independently for many centuries under the banners of the Arhat and the Bodhissattva ideals. An Arhat is a saint who has attained the fourth and highest stage attainable by a Hinayana Buddhist, has already freed himself from all craving and rebirth, and is worthy of receiving respect and offering. In early Buddhism the Buddha was called Arhat, but after the rise of Mahayana, Arhat was limited to the saints of Hinayana Buddhism.

In early Buddhism in Hinayana or Theravada, the term Bodhisattva designates a Buddha-to-be, or Sakyamuni Buddha's state prior to becoming a Buddha, who had been practicing austerities before his emancipation or before being born on earth. The tales of Sakyamuni Buddha's previous lives were elaborated and integrated into *Jataka* stories in the Pali literature.

In Mahayana Buddhism, the Bodhisattva aims at the enlightenment of all beings without distinction of self and not-self, whereas the Arhat aims only at self-enlightenment. Also in Mahayana Buddhism the quality of compassion is emphasized equally with wisdom, whereas Hinayana emphasized wisdom more than compassion. Thus, in Mahayana Buddhism, according to such texts as the *Prajnaparamita-sutra* or the Lotus sutra, the Bodhisattva refers to one who practices a series of six perfections (Skt. *paramita* refers to the perfecting practices), strives to attain enlightenment, vows to save all sentient beings, and works with compassion for suffering beings. The six perfections are: offerings or donation (*dana*), keeping precepts (*sila*), forbearance (*ksanti*), effort (*virya*), meditation (*dhyana*), and wisdom (*prajna*). Later some texts such as the *Dasabhumika-sutra* add four more perfections: skill in means (*upaya*), resolution (*pranidhana*), strength (*bala*) and knowledge (*jnana*).

BUDDHIST BIOETHICS

Bioethics and the Japanese Buddhist Movement

With advances in medical technology, and the fact that 70% of people die in the hospital, various Buddhist sects began in 1980 to study the effects of advances in medical technology upon problems of life and death, even though for Buddhism the final moments of a dying individual should be concentrated upon obtaining a peaceful mind. The Kyoto Young Buddhist Society was established in 1984, and the Death Education Seminar was started in Tokyo in the same year. The Kyoto Young Buddhist Society began holding annual conferences in 1988. On the institutional level, the Jodo Shin denomination Honganji-ha established the Practical Vihara Forum in 1986, and the Consolidated Buddhist Institute in 1990, both to deal with bioethical problems.

Problems such as the prevalence of deaths in the hospital, new reproductive techniques (artificial insemination with donated semen, in vitro fertilization and egg transfer), abortion, euthanasia, organ donation, and organ transplantation create a need for a more analytical view of life and death within Buddhism than Buddhism's traditional principles make available.

On the academic level, the Japanese Association of Indian and Buddhist Studies organized the Committee for Inquiry into Brain Death and Organ Transplantation in July 1988. They issued their opinions on six topics in September 1990. First, the committee thought that the issues of brain death and organ transplantation should be resolved by a relationship between donor and recipient that is based upon the Buddhist principle of respect for life. Second, they recognized that brain death is a critical step that inevitably leads to the death of all body cells, but ask the medical profession to reach a consensus about what it is that constitutes an individual's death. Third, they recognized that the donor, in offering his own organs to the suffering patient, is manifesting sublimity of spirit, and that the recipient should feel pain of heart upon receiving these organs which he needs for his own life. In this regard, they asked that medical doctors be very careful about what prudent medical practice dictates. Fourth, they noted that transplantation and the development of artificial organs should be recognized simply as means to an appropriate end, not as ends in themselves. Fifth, they said that organs should not be regarded simply as parts of the body. To recognize the sublimity of spirit in which organs are offered, laws are needed to prevent organs being bought and sold or secured by compulsion. Sixth, they recognized that the most urgent and important task for Buddhist medicine is to establish the *Vihara*, the Buddhist Hospice, to deal with terminal care.

The Application of Buddhist Principles to Various Current Concerns in Bioethics

1. *Abortion*

Statistics show that in Japan there are more than 500,000 abortions per year. The Eugenic Protection Law of 1948 legalized abortion. Although there was sentiment against the law within various established Buddhist sects, especially the Japan Women's Buddhist Federation, Buddhists in general have not voiced effective opposition to legalized abortion. The law permits abortion for eugenic and economical reasons. The clause stating eugenic reasons, revised in 1990 and implemented in 1991, permits abortion when the mother's life is in danger, and at any time within 22 weeks of pregnancy.

This law obviously contravenes, at least in spirit, the Buddhist view that life begins at the moment of conception. However, inasmuch as the principle of respect for life is the foundation of Buddhist ethics, Buddhism might be able to accommodate abortion if the mother's life is in danger otherwise.

2. *Active Euthanasia*

Buddhism absolutely cannot recognize active euthanasia, a form of mercy killing. Unlike suicide, which is an act of willfully taking one's own life, and may be justifiable if it is in fact self-immolation (see below), euthanasia is unacceptable, since the person is incapable of judging for himself. Even if an invalid begs to be killed for mercy's sake, he is under such extreme physical and emotional pain that his plea may not be genuine. It is a human instinct to want to live, and in actuality man tries to live for even an extra day. Therefore, the act of taking life is considered to be a blasphemy against man.

The Buddhist concept of compassion leaves no room to argue for active euthanasia. The fundamental Buddhist attitude underlying compassion is that a man must respect his own life, as well as the lives of others. In other words, Buddhism does not differentiate one's own self from others. Consequently, Buddhism forbids active euthanasia as well as suicide.

Self-immolation, however, is a spiritual ideal to be practiced as a *Bodhisattva*. It is an ideal best conveyed by parables. First, there is the famous story of self-immolation presented in the *Sessen doji*, which describes the past life of the Buddha. According to this work, Sakra, one of the two tutelary gods of Buddhism, transformed his body into a devil and approached a boy practicing Buddhism in the snow-covered Himalayas in order to examine his mind. The devil then started to hum the teaching of the Buddha. The boy listened carefully and greatly rejoiced. He asked the devil to teach him the entire teachings of Buddha and, in exchange, offered his body as food for the devil. The boy then engraved the teachings on a stone, a wall, and a tree, and jumped from the top of a cliff, aiming at the mouth of the devil. Immediately after jumping, however, the devil transformed his body back into its original figure. He then received the boy, put him down on the earth safely, and greatly respected him. This story is described in Volume 14 of *Daihatsu-Nehan-Gye* (the *Mahaparinirvana-sutra)*, and was introduced into the texts of the primary schools in Japan before World War II. Also, the *Gengu-Kyo* (the Teaching on the Wise and the Fool) describes the good king who offered his own eyes to an old blind person, even though he knew this was the request of an immoral king who governed another country.

The crucial point is that self-immolation, in contrast to both suicide and euthanasia, is the practice not of ending one's own or another's life, but of giving one's self over and merging one's self into the transiently real. Self-immolation is traditionally practiced by starving or drowning, as well as by burning. The *Mahaprajnaparamitopadesasastra* categorizes giving into three classes, and regards self-immolation as the highest human act.

3. *Brain Death and Organ Transplantation*

These issues became urgent in 1989 when Japan's Welfare Ministry established a panel to investigate brain death and organ transplantation. By the end of 1990, and prior to the implementation of an administrative decree establishing brain death, 10 children had already been transplanted with livers from their parents. According to a survey in the newspaper *Yomiuri* in December 1990, 51% of the Japanese approved of brain death criteria for determining death. In light of these developments, it is worthwhile to consider the relevance of Buddhism's ethical principles.

Consider first the conceptions of life and death. The chapter of the *Yakka-samyutta* in *Samyutta Nikaya* says that after conception, there exists *kalala*. According to Monier's *A Sanskrit-English Dictionary*, *Kalala* is "the embryo a short time after conception". *Kalala* grows into a small lump, namely *abbuda*, *pesi*, *ghana* and *pasakka*, from which hair, nail, and parts of the body grow ([5], Vol. 1, p. 206). From this time on, according to *Abhidharmakosa*, life is present. On the other hand, death is defined, according to *Mahavedalla-sutta*, as the moment when the three elements, namely, duration of life (*ayu*), fire or hot (*usna*), and consciousness (*vinnana*), leave the body ([6]; [4], Vol. 1, p. 296). These three elements are then regarded as *kalala*. Obviously, the Buddhist view of life and death is holistic, while the scientific and medical view is almost entirely analytical.

Buddhist ethics does not stress bodily integrity. According to Buddhism, our body consists of the five *skandhas* (aggregates), i.e., form, perception, conception, volition, and consciousness. Death scatters these five aggregates.

With respect to organ transplantation, the relevance of Buddhist ethical ideals must teased out. On the one hand, the Buddhist teaching of the "oneness of birth and death" counts against organ transplantation, because the psychological attitude behind organ transplantation tends to deny the transitory nature of both life and death. Moreover, the Buddhist ideal of giving, of which self-immolation is the epiphany, does not pertain to one's body.

On the other hand, the ideal of giving can in theory support the donation of one's organs to another. However, there is a condition for this support. Monks are supposed to chant the following verse whenever they receive offerings: "There should be no adherence to a donor, a donee, or a donation". Thus, although the Buddhist ideal of giving supports the donation of an organ, it requires that the recipient neither desire the donor's death, nor desire to prolong his own life.

In the author's judgment, the existence of both support for and opposition to organ transplantation can be attributed to the Buddhist notion of *Taikiseppo* (the skillful explanation of the teachings of Buddha such that the same teaching is expressed in various ways, depending upon the capacities and circumstances of the listeners). From its earliest period, Buddhism has functioned as a situational ethics, providing guidance for people in various time, places, and circumstances. This function has often led to equivocal moral conduct. For example, Japan imports corneas for corneal transplantation from the Buddhist country Sri Lanka. Nonetheless, the Buddhist conception of relativity, "to live for others and to contribute to others' lives", seems to support the practice of organ transplantation in the right spirit, namely, to help people who are in need.

Brain death, in contrast to organ transplantation, seems unequivocally opposed to Buddhist ideals. For brain death regards the body simply as an assemblage of organs, of which the brain is the critical one. Thus, if the brain dies, one is to disregard the fact that other organs are still functioning, and declare the whole body and the whole person dead. In this respect, the cycle of birth and death is broken; one is permitted to explant particular organs whose cells are alive from a body whose brain is dead. This way of thinking is incompatible with the Buddhist ideal of the "oneness of birth and death".

CONCLUSION

It is clear that the non-analytical nature of Buddhist ideals is often not very helpful in engaging particular bioethical problems. It is important, however, in thinking about these problems to adhere to the basic Buddhist principle of seeking the Right Way. This Way will differ according to each individual's Buddhist conscience. Moreover, it may be difficult to follow in many circumstances. One must simply accept the difficulty as his *karma*; this is what the spirit of compassion is for. Irrespective of social status, race, or ability, man should accept his *karma* and endeavor in all circumstances to strive for universal harmony.

Taisho University
Tokyo, Japan

BIBLIOGRAPHY

1. *Malunkyaputta-sutta*: 1948, in V. Trenkner and R. Chalmers (eds.), *Majjhima Nikaya*, No. 63, Vol. 1, PTS (Pali Text Society of London), London.
2. *Dammapada*, no. 183: 1914, in S. Sumangala Thera (ed.), *Khuddaka Nikaya*, PTS, London, 1914; also in T.W. Rhys Davids and J.E. Carpenter (eds.), *Digha Nikaya* Vol. II, PTS, London, 1903.

3. *Dhammacakkapavaitana-sutita*: 1969, in W. Oldanberg (ed.), *Vinaya, Mahavagga* 1-6, PTS, London.
4. *Majjhima Nikaya*: 1961, Ak. Warder (ed.), 2 rev. ed., PTS, London.
5. *Samyutta Nikaya*: 1898, in I. Feer (ed.), PTS, London.
6. *Mahavedalla-sutta*: 1961, in [4].

JUDITH A. GRANBOIS AND DAVID H. SMITH

THE ANGLICAN COMMUNION AND BIOETHICS[1]

The Anglican Communion is a loose confederation of churches whose historical roots can be traced to the Church of England. The various branches function autonomously; no branch can impose its will on any other. The Archbishop of Canterbury serves a symbolic function as a spiritual leader, but he has no power over the Anglican Communion outside the Church of England.

The lack of a central authority presented logistical challenges to the conduct of research for this report. We began by writing to the Provincials (executive officers) of all the churches in the Anglican Communion throughout the world, asking for any statements or resolutions their churches had generated on issues of biomedical ethics. Many branches (Canada, New Zealand, Australia, Hong Kong) never responded, which does not, of course, mean that they are not interested or active in studying these issues. Other branches, particularly in the Third World (e.g., Barbados, Brazil, Tanzania), stated that they had not addressed any of the issues we mentioned. Many bioethical issues in the First World (e.g., treatment of handicapped newborns, surrogacy, in vitro fertilization) require technology that simply is not widely available elsewhere. Many Anglican churches in the Third World focus primarily on more basic missionary activity, and it is important to note that the Anglican Communion, in which Caucasians are now the minority, is experiencing its most rapid growth there.

Every ten years all the bishops of the Anglican Communion are invited to convene for The Lambeth Conference, where diverse topics are debated and discussed. The Lambeth Conference has some moral authority, but no power to impose its will or views; it is not a "college of cardinals". Lambeth receives reports, debates resolutions, and issues pastoral letters on topics that are of general interest and concern; the documents from the most recent conference were compiled in *The Truth Shall Make You Free: The Lambeth Conference 1988* [1].

Lambeth produced no reports, resolutions, or pastoral letters that focused specifically on issues of biomedical ethics. However, the conference report offers some interesting and relevant insights.

One germane topic is the understanding of health and healing articulated in the report, which is discussed in more detail below. Another is the report's cautionary tone in its references to medical technology:

Increasingly, human invention has made possible achievements that were previously inconceivable. ...The potential is there for all to be freed from many of the ills which have, in the past, limited human life. We can look forward to the removal of other obstacles that impede human welfare. All these achievements give hope on the one hand, hope of what can be accomplished in the future; but they also bring with them a fear that their misuse could be destructive, and an uncertainty about our ability to handle what our science has created ([1], p. 156).

Our information about the Episcopal Church in the United States of America (ECUSA) is the most complete. Here our sources of information include the journals of General Convention and the Blue Books, which collect reports from the church's commissions and other bodies in advance of the triennial General Convention.

In the American Episcopal Church, biomedical issues have been assigned to a study commission. This body has undergone an evolutionary process over the years, manifested by changes in its title. (Until 1988, it was the Standing Commission on Human Affairs and Health.) The commission's agenda for the 1986-88 triennium included not only bioethics issues but also human sexuality, marriage, and institutional racism. The Commission has no staff, and its annual budget for the last triennium never exceeded $13,500. Its 1988 report acknowledged the difficulties of fulfilling its ambitious mission with these limited resources:

> The Commission on Human Affairs and Health has addressed many issues in bioethics, all of which are complex and in some cases fast-changing. It needs to be said that time did not allow us to focus on these issues except in the most cursory way. The areas addressed in this report include infertility, genetic experimentation, advance directives (living wills) and organ transplants ([16], p. 157).

> [W]e offer this caution: This commission's agenda is already so broad that it is difficult to address adequately all the matters referred to it ([16], p. 161).

Perhaps in response to these concerns, the commission was divided into two separate bodies in 1988, the Standing Commission on Human Affairs and the Standing Commission on Health.

UNDERSTANDINGS OF HEALTH AND HEALING

It is instructive to review the theological context in which biomedical issues are studied in the Anglican Communion. The 1988 Lambeth Conference articulated its conception as follows:

> In the widest sense of the words, healing and salvation are two ways of describing the work of Jesus Christ. Within the work of Christ's salvation the healing of the sick holds a significant place because by it he both demonstrated the love of God and proclaimed that the Kingdom had come in his person ([1], p. 47).

The report refers to the affirmation of the church's ministry of healing by the Lambeth Conference of 1920, which emphasized teaching, cooperation with medical professionals, and prayer. The 1988 report also quotes Archbishop William Temple, who established the Churches' Council for Health and Healing in England in 1944:

The Church is called to exercise a ministry of healing to the sick whatever their condition of mind and body. Such a ministry of love aims: (1) to restore the sense of relationship with God and the community; (2) to affirm the body's natural powers of healing; (3) to use all medical knowledge and skills to assist restoration or relieve pain; (4) to establish peace as we acknowledge that we are mortal and yet are born for eternity ([1], pp. 47-48).

The 1988 report of the American church's Commission on Human Affairs and Health echoes this language and goes on to invoke arguments of justice in affirming the church's role in the ministry of healing:

The commission has struggled with...[a] profoundly important question: what is health? Any notion of health that is to be taken seriously by Christians must include those convictions and conclusions that God, working through human experience, has revealed to be healthy for us. Salvation through Christ, or wholeness in Christ, is the larger conception of what health means to Christians. We never achieve the fullness of salvation, health or wholeness in our lifetime in this world that is, of course, the only world we have. We join all other persons in asserting that there are fundamental rights to physical and mental health to which human commitment and ingenuity must be dedicated. We, therefore, make common cause in these things with all of the people of the earth ([16], p. 139).

INFERTILITY AND ASSISTED REPRODUCTION

Issues of assisted reproduction have occupied several branches of the Anglican Communion, including the Scottish Episcopal Church. The "Report of the Working Party On Human Infertility and Embryo Research" was produced to "respond from within the Scottish Episcopal Church to a request from the Scottish Home and Health Department to comment on proposed legislation on Human Infertility and Embryo Research". The authors underscored the fact that they were speaking from *within* their church and not *for* their church:

We recognised throughout that we were dealing with intimate and personal issues which affect national life. We were privileged that from our Christian perspective we could make an offering to those whose task will be to draft legislation. This Christian perspective involves a pastoral framework of loving care held within the tradition of a Biblically based Church, nourished within an ordered (and sacramental) structure of worship and spirituality....We considered that absolutist views in this area do not take into account the complexity of individual situations nor the tragedy that may also be present....Our intention has therefore been to work within a framework of pastoral care that attends to the voice of Scripture and tradition and the developing witness of the Church in an evolving world ([10], p. 7).

Dated May 31, 1987, the report summarizes the issues and includes theological reflections and recommendations about forthcoming legislation. It touches upon a wide variety of topics, including infertility, artificial insemination by donor (AID), in vitro fertilization, gamete intrafallopian transfer, surrogacy, access to genetic records for those born through sperm or ovum donation, anonymity of donors, embryo research, storage of eggs and embryos, trans-species fertilization, and contraception.

The report strongly encourages the creation and empowerment of a licensing authority to oversee the many complex moral, legal, and financial questions raised around these issues. It asserts that, "[n]o practice with regard to Human Infertility Services, Embryo Research or Surrogacy Services should be permissible without license" ([10], p. 13). The licensing body should be empowered to create a code of practice, to inspect and license any facility or person "wishing to engage in research on in vitro fertilization and embryo research" and any surrogacy agency, and to "keep the public as informed as possible about the nature of its work and of the results of its investigations." ([10], p. 14). The report also encouraged legislation that would give such a licensing authority appropriate power to carry out its duties, with accountability to the Secretary of State.

The report also states that "[n]on-directive counselling by trained and qualified counsellors of those involved should be a statutory requirement of infertility and surrogacy services", but it specifies that "[t]here should be *no* assessment for suitability for prospective parents seeking AID/Donation" ([10], p. 15). The birth certificates of children born as a result of these arrangements should be "identical in every respect to the standard Birth Certificate" ([10], p. 16); any children so born should be accorded full legitimacy; the parents "should be recorded as those in whose home the infant is to be nurtured and brought up" ([10], p. 15). Only minimal records should be kept of the "information relevant to AID/Egg donation" ([10], p. 16), excluding identification of the donor; hence children would not have access to information about their genetic origins.

Authors of the report argued that embryos should be stored no longer than ten years. "On the death of one partner the right of use or disposal passes to the surviving partner", but "a child born by such means should be disregarded for succession of inheritance from the father after the father's death" ([10], p. 17).

Issues around surrogacy also commanded significant attention:

> The welfare of the mother-and-child viewed as a fundamental unit would be the guiding factor in resolving any problem and dilemma that might subsequently present itself in a surrogacy arrangement....The commissioning parents and the surrogate mother should be free from prosecution....Advertising of, or for surrogacy services should be prohibited ([10], p. 19).

The authors conclude that no woman should enter into a surrogacy arrangement without the written consent of her husband, and they comment on the provisions of surrogacy arrangements:

> In a surrogacy agreement or contract there should be no requirement for the surrogate mother to produce a child without mental or physical handicap or genetic disorder. When a child is born with such a handicap and the commissioning couple refuse to take the child, parental responsibility remains with the mother who gave birth and her spouse ([10], p. 20).

An "honest broker", a disinterested third party, should be appointed to "hold, and dispose of, monies pertaining to the surrogacy arrangement or contract" ([10], p. 20). The report also states that "[s]perm donations, egg donations or surrogacy arrangements shall result in no more than six children per donor" ([10], p. 20).

The Working Party on Human Infertility and Embryo Research concluded the theological reflection in its report by calling attention to other pastoral concerns:

> We also recognised a need for the Church's sympathetic and pastoral care for those engaged in infertility services. On the whole we are reassured by the sincerity of research workers. Whilst some researchers and diagnosticians are concerned with advancement of personal prestige, we believe most to be honourable and caring ([10], p. 9).
>
> We considered that even those who cannot agree with infertility research (etc.) should moderate their criticism of it as they may well be needed for help and counsel. Expressions of criticism will inhibit the practice of caring and Christian help. This is the case with respect to seemingly obvious cases of good and right, as well as to those of bad and wrong. In some situations the only choice will be between lesser or greater evils. In all of these the Church must continue to enable God's love to flow ([10], p. 9).

Issues of assisted reproduction also concerned the Church in Wales. "Human Infertility Services and Embryology: Some Ethical Issues" is a report to the Governing Body of the Church in Wales, produced in April 1989 by the Bio-Ethics Working Group, Division for Social Responsibility.

In addition to its 29-page report, the working group also published a five-page pamphlet, suitable for wider distribution. The pamphlet briefly summarizes the longer report and discusses the issues in a question-and-answer format. It provides information about the various techniques of assisted reproduction, and discusses some of the ethical questions the Working Group considered important for persons who are contemplating infertility services, and for all interested Christians.

According to the report,

> Human infertility must engage the Christian Conscience. It poses, not only biological problems, but also religious problems, spiritual problems....If a couple has a moral duty to "be fruitful and multiply," then society likewise has a moral duty to assist them towards that end ([4], p. 4).

Techniques such as artificial insemination by the husband, artificial insemination by donor, in vitro fertilization, embryo transfer, gamete intrafallopian transfer, zygote intrafallopian transfer, and surrogate motherhood, however, not only give hope to those who are infertile; they also raise moral questions that must be addressed.

Do people have a "right" to have children? If, the report argues, the claim is asserted that individuals are

> not only not to be prevented from having children by, for example, compulsory sterilisation, but

also to be provided with artificial reproductive techniques if necessary, then this in itself presupposes that there exists some person or group of persons whose responsibility it is to provide these services. If the cost is to be met by the public purse then this claim has to compete with others which are making demands on the limited resources available ([4], p. 9).

Because infertility is not a disease, the authors ask, is it within the medical domain? Will publicizing techniques of assisted reproduction "make it harder for people to accept childlessness and to channel their energies elsewhere" ([4], p. 9)?

Beyond the level of individual choice, reproductive technology raises various societal issues. Is marriage the only context in which fertility questions should be considered? As the report states, "those who argue for a basic human right to have children for all individuals may argue for the provision of reproductive services not only to married couples, but also to unmarried heterosexuals, homosexuals and lesbians, and single people who simply prefer to have a child in this way" ([4], p. 10). Does society have a right to determine the type of family that may produce a child through the use of these techniques? What are the interests of the child? What is the possible impact upon society at large? The report raises these questions in the interest of debate; it offers no specific policy statements.

The Working Group does, however, articulate ethical objections to AIH, AID, and IVF. They argue that it is possible to object to artificial insemination, first, because it separates "the procreative and unitive functions of sexual intercourse" ([2], p. 3), and, second, because it involves masturbation.

The Working Group grants that artificial insemination by donor or egg donation can be perceived as stepping outside the marital vows of fidelity. "However, since the *intention* is...to have a child within the family, it may be seen rather as an affirmation of the family, and so an acceptable practice" ([2], p. 4). Should donors ever be compensated? Never, the pamphlet asserts. Parents are encouraged to inform their child about his or her genetic origins, rather than pretending biological parenthood. Such openness is in the child's best interest.

In vitro fertilization can raise some of the same ethical questions as donation of egg or sperm by someone other than the spouse. Again, the intention of the couple is seen as crucial to the morality of the situation. The authors acknowledge that the procedure of IVF often results in "spare" embryos and express concern about what happens to them. The report encourages research that would perfect the technique of storing ova, hence decreasing the need to store spare embryos.

The Working Group also expresses concern about the *cost* of IVF. Is it appropriate to expect the National Health Service, already considerably strained by the provision of necessary medical services, to offer this expensive (and frequently unsuccessful) service? The report acknowledges that other moral issues are raised if only those who can afford to go to private clinics have access to the procedure.

Of the various techniques of assisted reproduction, the Episcopal Church in the United States has focused primarily on in vitro fertilization (IVF), artificial insemination (by donor or by husband), and surrogate motherhood. It has sanctioned technological assistance for couples who are using their own gametes, but it has frowned upon if not condemned the introduction of a third person's gametes.

The 1988 General Convention adopted a Resolution commending clinicians, ethicists, lawyers, and others who "study with care the spiritual and moral dilemmas of surrogate parenthood"; recommending that those who are considering surrogate parenthood consult clergy or other counselors about the spiritual and psychological issues; encouraging study and discussion of the question; and requesting each bishop to identify and train informed clergy and counselors ([12], pp. 634-635). The Resolution offered no substantive guidance on any of the profound and troubling issues surrounding these practices. What insight is available into the current thinking on the subject must be provided by the reports of the Standing Commission on Human Affairs and Health.

The Standing Commission addressed several germane issues in its 1988 report under the heading of infertility [16]. The report endorsed artificial insemination by husband (AIH), but withheld its approval of reproductive strategies that introduce a third person into the marital dyad (e.g., artificial insemination by donor and surrogate motherhood). It affirmed fertility-enhancing drugs, with the warnings that couples should be prepared for therapeutic failure and the increased risks of tubal pregnancy and maternal and fetal death. On the other hand, the report warned couples to prepare financially and emotionally for the possibility of therapeutic success that might produce multiple ovulations and multiple births.

One underpinning for these essentially favorable conclusions is the assertion that technology is not wrong solely because it may be in some sense "unnatural". For example, the 1982 Report of the Standing Commission on Human Affairs and Health noted that some voices had been raised in objection to IVF "on the basis that one is grossly interfering in the reproduction process, a process or physiological function which some presume to have special protection" ([14], p. 134). The Commission believed this concern was unwarranted, however. Again, in 1988, the Commission observed, "Anglican and Protestant ethics generally...affirm AIH as a means to help a married couple conceive children. In this case the good of conception in marriage justifies these means to make conception possible" ([16], p. 158).

On the other hand, as the citations above suggest, the use of semen or ova donated by a third party raises troubling issues. The Reports of the Commission reflect an evolutionary process and an ultimate hardening of opinion against these practices.

For example, the 1973 report noted that "[t]here is serious question of the propriety of the use of semen derived from a donor in situations where the concept of the family is in question to begin with or where it might be raised later" ([11], p. 595). In contrast, however, the 1982 report expressly affirmed "the value of surrogate parenting through the means of the anonymous male donor", because "the wholeness of the marital union brought about through wanted children must be valued above the risk" ([14], p. 133). The softening toward AID continued in 1985. That year's report claimed that the church has not opposed AI, given that the attitude of most couples toward procreation is loving and unselfish. But the report ringingly rejected the interjection of money into the process, and deplored the "congruence of consumerist values in personal relationships with technical advances in modern obstetrics" ([15], p. 141) that enables the sale of embryos: "[T]o the extent that such an embryo is a person, a human being may soon be for sale. Can the Church condone the buying and selling of the flesh into which Christ came down from heaven?" ([15], p. 141).

Flirtation with support for use of donor gametes ended in 1988. Noting that "[s]erious moral objections are raised by Christian ethicists about fertility enhancement by a third person" ([16], p. 158), the Commission concluded that it "cannot affirm AID as a moral option for members of this Church" ([16], p. 159).

In contrast to its wavering on AID, the Commission has consistently expressed grave reservations about surrogate motherhood. As early as 1973, the Joint Commission on the Church in Human Affairs asserted that "the use of a host uterus to obtain the gestation of a child outside the family almost necessarily involves the treatment of the second woman as a mere object" ([11], p. 595).

Concern with the exploitation of persons is a recurring theme in the church's deliberations on this topic. In 1982, the Commission on Human Affairs and Health distinguished between artificial insemination and surrogate motherhood, noting that the differences are tacitly acknowledged by the prices charged for the two services. The sperm donor's identity is unknown, and he forms no emotional attachment to the child who may be conceived, in sharp contrast to the experience of the surrogate mother [14].

Drawing an analogy between surrogacy and prostitution in its 1985 report, the Commission observes that "[s]urrogate motherhood...needs to be examined in incarnational terms" ([15], p. 141). In the Commission's view, surrogacy differs fundamentally from artificial insemination in terms of its profound emotional impact not only on the surrogate mother, but also on her other children and the adoptive parents. Granting the positive values of providing a child to a childless couple (absent genetic problems or manipulative behavior by the parties to the agreement), the Commission appeals to various risks associated with surrogacy. The arrangement gives the biological parent of the couple a "very powerful lever" over the adopting parent, for example, in situations of crisis or stress. The child conceived through surrogacy also has a weapon against the adoptive parent [15].

The Commission also points to the primary emphasis in the *Book of*

Common Prayer on the relationship of husband and wife, their "mutual joy", and "the help and comfort given one another". Here, the procreation of children is a secondary issue, "when it is God's will" [15]. The 1988 statement casts the matter in more theological terms, noting that "moral issues are usually discussed in American medicine and ethics outside any specifically Christian vision of marriage, parenting, and sexuality...due to the widespread secularization and indeed, secularism, of Western culture....[The Commission] cannot affirm surrogate motherhood as a moral option for members of this Church" ([16], p. 159).

Finally, and perhaps centrally to the whole tradition of its discussion of the issues surrounding assisted reproduction, the Commission appeals to the relational nature of sexuality as it is understood by the church. Thus, its 1982 report rejects the use of technology to provide a child for a single person as an action that

> appears to us to be in the mainstream of American narcissism and self-indulgence. The child is brought to a single person household for reasons of personal self-satisfaction and achievement rather than for the completion of God's holy union, and is to be condemned ([14], p. 133).

The Commission reaffirms this conclusion in its 1988 report, noting that a single woman conceiving her "own" child through AID or IVF is

> deliberately causing a child to suffer the deprivation and risks of having only one known biological parent....These techniques place a primacy on having "our baby" or "my baby" over the care and nurturing of children. Is such a biologically biased view of parenting morally justified in the light of existing children needing adoption ([16], p. 159)?

ABORTION AND RESEARCH ON FETUSES AND EMBRYOS

The ethical dilemmas presented by abortion have preoccupied several branches of the Anglican Communion. The Church of England, for example, is in conflict over this issue. The Abortion Act of 1967 allowed abortion in the United Kingdom up to the twenty-eighth week of gestation, but the appropriate gestational age limit is only one disputed question.

The Church of England has also extensively debated the issue of embryo research. The Rev. Professor John Polkinghorne produced a report (referred to as "The Polkinghorne Report") on the research use of aborted fetuses that drew comment from the Church in Wales and the pro-life Society for the Protection of Unborn Children (SPUC), as well as from persons involved in the drafting of legislation for Parliament. Legislation referring to human fertilization and embryology sometimes included attempts to limit the Abortion Act, a connection that generated both support and criticism from differing quarters of the Church of England.

Because some bishops in the Church of England have seats in the House of Lords, these various pieces of legislation are hotly debated in church publications, with the views of some Church of England clergy well represented. For example, when the issue was debated in February 1990 and the House of Lords "opted to allow controlled experiments on embryos to continue" ([6], p. 2), the Archbishop of York and the Bishop of London took opposite sides in the debate. Debate in both the House of Commons and the House of Lords was generously reported in *The Church Times*. At the same time an editorial appeared stating that "Parliament is not an ideal body to decide these questions" of embryo research and abortion. "The balance being discussed in these questions", the editorial concluded, "is a fine and complicated one, and not to be solely determined by an appeal to moral absolutes" ([9], p. 10).

The Scottish Working Party also addressed the issue of embryo research. "The unlawful use of an in vitro embryo should in itself constitute a criminal offense" ([10], p. 22). They acknowledged that "the fourteen day limit on embryo research is justifiable...if the research thus enabled better ante-natal diagnosis and improved infertility treatment" ([10], p. 22). They also noted that such research might result in more efficient methods of contraception and thus have a significant impact on global population trends.

Issues of embryo research have also aroused concern in the Church in Wales. The Bio-Ethics Working Group's report concluded that it is more acceptable to conduct research on gametes than on embryos created in vitro. The authors argued that in no circumstances should embryos be "created" for research purposes; the only acceptable "end" for in vitro fertilization is the intention to implant. Determination of sex or "genetic purity" should never be criteria for destruction of an embryo [4].

The Working Group examined the question of research on embryos in some detail:

> Is it ethically permissible to carry out experiments on these 'spare embryos', not for their own good (since in the process they will be killed), but to further the interests of medical research (especially in the field of genetically transmitted diseases, or cancer research)? Is it, indeed, ethically justifiable simply to allow surplus embryos to die in the laboratory ([4], p. 5)?
>
> ...[A] cluster of cells cannot have the same status as a growing organism. There is a moral as well as a material discontinuity between the pre-embryo, the embryo and the neonate; between the 'primitive streak' stage and birth, there are long processes of defining identity, and in developing appropriate ethics such facts have to be recognised and their implications assessed. There are, indeed, no simple, clear-cut answers to the problems posed by genetic manipulation and embryo research. But our concern for humanity, while it does not justify a practical absolutism in this area, does impel us to a stance of caution, appreciating the powers which God has given us as co-creators in his cosmic exercise of loving creation, but scrutinizing the motives of all those involved, and concerned that all infertility treatments be set firmly in the context of living interpersonal relationships, and that the products of conception be treated with the respect due to human matter ([4], p. 6).

In the United States, the General Convention passed resolutions on abortion in 1967, 1976, 1979, 1982, 1985, and 1988.[2] The 1967 Resolution, which predated the Supreme Court's decision legalizing abortion in *Roe v. Wade*

(1973), supported legal reform to permit abortions to protect the mother's physical health, when mental or physical deformity is diagnosed in the fetus, or when the pregnancy results from rape or incest. The 1976 Resolution changed its focus to individuals considering abortion, stressing the moral seriousness of reproduction and affirming responsible birth control (excluding "abortions for convenience") as a morally legitimate means of limiting family size. Episcopalians seeking abortion for reasons other than those enumerated in the 1967 Resolution and endorsed in 1976 were urged to seek the advice and counsel of a priest and, where appropriate, Penance; further, they were urged to explore "other preferable courses of action". The Resolution ended by expressing "unequivocal opposition to any legislation on the part of the national or state governments which would abridge or deny the right of individuals to reach informed decisions in this matter and to act on them" (quoted in [13]).

This position was essentially reaffirmed by the resolutions passed in 1976, 1979, and 1982. In 1988, however, the Standing Commission on Human Affairs and Health proposed a new resolution that represented a departure from the past. Novel elements included: (1) an assertion that all human life is sacred from "inception" until death; (2) the claim that "*congregations* have responsibility to assist members in becoming informed concerning the spiritual, physiological and psychological aspects of sex and sexuality", and an explicit statement of the Church's "obligation to help form the consciences of its members concerning this sacredness [of human life]"; (3) a reference to the formulation in the *Book of Common Prayer* that the birth of a child is a joyous and solemn occasion for families and the Christian community; and (4) an enumeration of reasons that would render abortion immoral: "as a means of birth control, family planning, sex selection, or any reason of mere convenience" (quoted in [13]).

David A. Scott points to several shifts in focus in the 1988 Resolution. First, it clearly addresses the Church and members of the Christian community. Second, it moves, in Scott's formulation, "from boundary to bias...from drawing boundaries between morally permissible and impermissible abortions to stressing the positive good of childbirth and the evil of abortion." Alternatives to abortion, such as adoption, are specified for the first time. Third, the focus shifts from advocating civil legislation legalizing abortion to urging law makers to respect individual moral conscience. The Resolutions between 1976 and 1982 expressed unequivocal opposition to legislation that would abridge the rights of individuals to act upon private decisions concerning abortion. In contrast, the 1988 Resolution addressed legislators as moral agents, asking them to "respect the moral conscience of citizens affected by legislation they pass" ([13], p. 9).

Scott criticizes the 1988 Resolution as inconsistent: If all human life is sacred, how can any abortions be justified? What abortions might be justified by "extreme situations"? (And, incidentally, how are "extreme situations" defined?) Further, he asserts that the Resolution is evasive. Noting the use of the vague term "inception" in place of "conception", he argues that the

Resolution fails to specify when human life acquires the attribute of sacredness. The Resolution fails to take account of the social, political, and cultural contexts in which abortion is considered, and it fails to provide guidance to medical caregivers or to women and men who confront the decision. Scott points to the Resolution's charging local congregations with the responsibility of informing their membership about these questions, arguing that this strategy effectively limits the whole church's role as a corporate body in forming the moral conscience of its members. Finally, he notes that the Resolution fails to appeal to the traditional standards of authority cited in Anglican moral teaching: scripture, tradition, and reason [13].

In its 1988 report, the Standing Commission on Human Affairs and Health (ECUSA) briefly discussed genetic engineering and gene splicing on three levels, noting that "the commission has not had time to reach conclusions regarding these levels and their attendant theological and moral aspects" ([16], p. 159). In the report's formulation, Level 1 includes study and experimentation on somatic human cells and cell cultures for the benefit of human health (e.g., production of insulin or interferons). The Commission endorsed this kind of research, provided that it is approved through peer review and intended for publication in appropriate scientific journals. However, the Commission excluded experimentation on human embryos from its endorsement, and expressed reservations about cryopreservation of human ova on the basis that the long-term effects have not been determined. The Commission declined to reach any conclusions about Level 2 (the development of genetic therapy) and Level 3 (the enhancement of human traits by genetic means), on the grounds that this work is still too experimental to permit any conclusions.

LIVING WILLS

The 1988 report of the Standing Commission on Human Affairs and Health (ECUSA) included a limited discussion of living wills. The report observed that advance directives are intended (1) to attest the individual's willingness to donate organs or tissues, and (2) to guide caregivers and survivors as to the individual's wishes regarding terminal care. The commission generally affirmed living wills, noting that "Christian values regard medical care and physical health as goods but not absolute goods or ends. Christian values tell us there are greater goods than material perdurance..." ([16], p. 160). The commission also noted that it had not addressed some troubling issues, including the difficult case of "permanent vegetative state":

> If we would not bury persons in a permanent vegetative state who are yet breathing on their own, is it morally right to end the breathing deliberately in order to bury them?...[Other issues to be addressed] are the distinction between acts of omission...and...commission..., the rightness of refusing life-saving treatment, truth telling, the distinction between the best interest standard and the substituted judgment standard in proxy decisions..., and the distinction between 'ordinary' and 'extraordinary' care ([16], p. 160).

ORGAN TRANSPLANTATION

The final issue addressed in the 1988 report of the Commission on Human Affairs and Health (ECUSA) is organ transplantation. The commission took up the question from two perspectives: the decision to donate an organ, which it found morally permissible, and the decision to be an organ recipient. On the latter point, the commission merely raised the issues of the moral limits on avoidance of death and the risk to donors. The commission warned that exceptional care must be taken to avoid coercion of living donors (e.g., through the institutional separation of surgical and organ retrieval teams) ([16], p. 161).

A special case concerns the transplantation of fetal tissue. The Bio-Ethics Working Group of the Church of Wales expressed deep concern about the use of human fetal tissue. Their report endorsed the guidelines created by the British Medical Association, making "an urgent appeal for action to make the B.M.A. guidelines statutory, enforceable in both the public and the private sector" ([4], p. 17).

The Polkinghorne Report also raised questions about fetal tissue transplants. For example, do those who receive such transplants have the right to know if the fetus was deliberately aborted or died from natural causes? The report recommends that such information not be divulged; the Society for the Protection of Unborn Children disagreed with that finding.

EQUITABLE ACCESS TO HEALTH CARE

"How healthy is the National Health Service?" is a question often asked in the United Kingdom. Widely discussed issues include the overall quality of medical care, preventive medicine, care of the dying, and long waiting lists for elective surgery. Some disquiet also is apparent about the quality of medical care available to the poor. The Christian perspective on the Welfare State was discussed in many articles in *The Church Times* throughout 1987 ([5] - [9]), e.g., when the General Synod of the Church of England "called on the country's politicians to ensure that major public services are available to everyone" ([5], p. 1). The high cost and problematic availability of medical care continues to be an issue, addressed in a 1989 editorial by Douglas Brown:

> Christian arguments for a political-legal right to health care must appeal to such moral principles as justice, equality, equity and fairness on the one hand and, on the other, love and charity that bears in mind that health needs are usually undeserved, randomly distributed, unpredictable and, for the person concerned, overridingly important ([3]).

The editorial also expresses concern that the original National Health Service ideal of "equal levels of treatment and amenities for all, regardless of income, status or education", is lost in the confusion of budgetary strains, and draws attention to the luxurious care that is available to those who can afford it.

In the American church, the Standing Commission on Human Affairs and

Health echoes these concerns. Its most recent statements on this issue relate to just distribution of economic resources more generally:

> [I]n the immediate world where most Episcopalians live, we see a recovering American economy and a dazzling picture of scientific wonders with brilliant technology....Too easily forgotten by those of us so blest, however, are the more than thirty-five million men, women, and children in this land who do not share these creaturely joys....Is the present operation of the American economic system as just and equitable as we can make it? Christian people must admit it is not. Does it take into consideration the social costs and consequences it creates, deliberately or otherwise?...Questions like these need to be...asked in light of the biblical narratives that shape our religious identity. In the Genesis creation stories, for example, human beings are clearly understood to be created primarily for relationship to God and to each other. They share God's image in their ability to shape the natural order, but, most of all, in their ability to love, to create a common life or community ([15], pp. 119-121).

Indiana University
Bloomington, Indiana, U.S.A.

NOTES

[1] We want to thank Mary J. Mail for her invaluable assistance in the research and writing of this report.
[2] For this discussion we are indebted to [13].

BIBLIOGRAPHY

1. Anglican Consultative Council: 1988, *The Truth Shall Make You Free: The Lambeth Conference 1988, The Reports, Resolutions and Pastoral Letters from the Bishops*, Church House Publishing, London.
2. Board of Mission, Division for Social Responsibility, Bio-Ethics Working Group: 1989, *D.S.R. Briefing. Human Infertility Services and Embryology: Some Ethical Issues, A Report to the Governing Body of the Church in Wales* (April 1989), Church in Wales Publications, Penarth.
3. Brown, D.: 1989, "As I See It: Health Challenge to the Churches", *Church Times*, March 31, p. 10.
4. Church in Wales Board of Mission, Division for Social Responsibility: 1989, *Human Infertility Services and Embryology: Some Ethical Issues, A Report to the Governing Body of the Church in Wales* (April 1989), Church in Wales Publications, Penarth.
5. *Church Times*: 1987: "Synod Plea to Politicians on Welfare", February 27, p. 1.
6. *Church Times*: 1990, "Bishops Voice Differing Views Over Embryo Experiments", February 16, p. 2.
7. *Church Times*: 1990, "Commons to Take Up Abortion Dispute Along

With Embryos", April 6, p. 3.
8. *Church Times*: 1990, "Commons Sets Term to Abortion and Embryo Research", April 27, p. 1.
9. *Church Times*: 1990, "Not A World Where Moral Absolutes Can Settle Everything", April 27, p. 10.
10. Gillies, R.A.: 1987, "Report of A Working Party on Human Infertility and Embryo Research", Scottish Episcopal Church (mimeograph), May 31.
11. *Journal of the General Convention of the Protestant Episcopal Church in the United States of America, 1973*: 1973, Episcopal Church Center, New York.
12. *Journal of the General Convention of the Protestant Episcopal Church in the United States of America, 1988*: 1988, Episcopal Church Center, New York.
13. Scott, D.A.: 1990, "Episcopal Ethical Teaching: Case Study Abortion", unpublished paper presented at meeting of Episcopal Ethicists on Moral Authority and Teaching in the Church, Seabury-Western Theological Seminary, Evanston, Illinois, November 9-10.
14. Standing Commission on Human Affairs and Health, The: 1982, "Report on Health", in *The Blue Book: Reports of the Committees, Commissions, Boards and Agencies of the General Convention of the Episcopal Church, 1982*, Episcopal Church Center, New York.
15. Standing Commission on Human Affairs and Health, The: 1985, "Report of the Commission", in *The Blue Book: Reports of the Committees, Commissions, Boards, and Agencies of the General Convention of the Episcopal Church, 1985*, Episcopal Church Center, New York.
16. Standing Commission on Human Affairs and Health, The: 1988, "The Commission's Report", in *The Blue Book: Reports of the Committees, Commissions, Boards, and Agencies of the General Convention of the Episcopal Church, 1988*, Episcopal Church Center, New York.
17. Turner, P.: forthcoming, "Social Advocacy As A Moral Issue in Itself, or What Ought The Churches To Say Or Not Say About *Webster v. Reproductive Health Services?*", *Journal of Religious Ethics*, forthcoming.

STANLEY SAMUEL HARAKAS

EASTERN ORTHODOX BIOETHICS

This bibliographical description of Eastern Orthodox Bioethics for the period 1988-1990 will begin with a background survey of some Eastern Orthodox bioethical thought, and then describe briefly some broad-based Eastern Orthodox reflections regarding bioethical issues during the past several decades. It will finally describe topically positions of some writers of the period under consideration. Thus, it seeks to provide an overview and context for future annual reports.

I. A BACKGROUND SURVEY

The discipline of bioethics, the beginning of which may perhaps be coterminous with the publication of *The Encyclopedia of Bioethics* [48], was, of course, not born in a vacuum. Discussion of bioethical issues has taken place throughout history, though without that specific designation.

While it is not correct to say that Eastern Orthodox Christianity is uninterested in ethical questions, it is correct to affirm that its approach is different in emphasis from that of traditional Roman Catholic and Protestant ethical concerns. Worship and liturgy, spiritual and mystical life, monastic *askesis* and polity, communal focus, pastoral care, doctrinal purity, and the cultivation of an ethos of love in the Christian life interest Eastern Christian tradition much more than formulations of abstract ethical constructs. Given this viewpoint, the Orthodox Church has not developed, in the main, complex rationalistic methodologies for bioethics. Those positions that it has developed are rooted in religious perspectives and not speculative reasoning.

Since Eastern Orthodox Christianity perceives itself to be in direct historical continuity with the undivided Christian Church of the first eight centuries of the Christian era, a "background survey" of Eastern Orthodox bioethical thought would logically require a bibliography covering nearly two millennia. Let it suffice to highlight one of the works that exercised great influence in the Orthodox Church during the first three decades of the present century. For almost forty years Basil (Vasileios) Antoniades taught at the Theological School of Halki, the chief center of education for the Hierarchy of the Ecumenical Patriarchate of Constantinople (Istanbul, Turkey) during that period.

Of his many writings, his two-volume *Handbook of Ethics According to Christ* [2] was issued in duplicated form originally and revised several times until its publication in printed book form in 1927. Antoniades was one of the first Orthodox writers to deal systematically with matters related to issues now grouped under the rubric of "bioethics".

Given the sacramental and communal emphasis of Eastern Orthodox Christianity, marriage and family values provide an important context for bioethical issues. Antoniades dealt with these under the rubric of "Love" ([2], sections 90-100). Under the heading *Autarchia* (Self-Governance) he dealt with issues such as "life and health" ([2], Sec. 40), "defense of life" ([2], Sec. 41), "suicide" ([2], Sec. 43), "abuse of life" ([2], Sec. 43), "sickness, death and euthanasia" ([1], Sec. 45), and "public health" ([2], Sec. 46). Under the heading "Development and Commitment to the New Life in Christ", he discussed issues of "diet" and "physical exercise" ([2], Secs. 27-28).

Of course, few of the most burning issues of our day had come into existence during the period in which he was writing, so he could not have addressed them. Nevertheless, his contribution is important for Eastern Orthodox ethical approaches to bioethical questions. Most significant was his attitude of openness to scientific developments, coupled with a critical evaluation of the scientific developments of his times as they impacted on the lives of persons. This "critical openness" is the greatest legacy Antoniades bequeathed to the Eastern Orthodox Ethical tradition.

Others writing on Orthodox Ethics in general, with some reference to issues of a general bioethical character, were Christos Androutsos [1], Nikodim Xristov and Ivan Pantchovski [50], John Karmires [37], and Panagiotes Demetropoulos [13].

In the period before 1950, two continuing concerns among Orthodox writers were abortion and conception control. The latter issue was raised and promoted by Serapheim Papakostas in Greece. In 1933 he published a book on the subject [45], in which he used theological, philosophical, medical, and demographic arguments to condemn all forms of birth control. Papakostas subsequently authored the text of an Encyclical by the Holy Synod of Greece on the topic, issued in 1937, which condemned birth control practices. It was supported in subsequent works by Meletios Galanopoulos [17] and Gabriel Dionysiatou [15] in the mid-1950s. Almost four decades were to pass before Papakostas' book was to be seriously challenged in Orthodox circles.

Part of the reason justifying this change was a gradual reappreciation of Eastern Orthodox understandings of the sacrament of matrimony, freeing it from certain western Christian influences. In France, a well-articulated approach illumined by Orthodox Trinitarian theology and the Palamite focus on salvation in sacramental and participatory categories was applied to marriage by Paul Evdokimov in his book *The Sacrament of Love* [16]. Jerome Cotsonis, who subsequently became Archbishop of Athens, introduced this perspective to Greece and tied these rediscovered understandings of marriage even more specifically to moral questions. His study was translated and published subsequently in English in 1964 [12]. In 1968 Meletios Galanopoulos, a bishop of the Church of Greece, published a book of 121 pages in which he discussed birth control measures, abortion, and euthanasia [18]. The book reflected traditional attitudes of opposition to these practices.

Little was written about other issues more commonly associated with bioethical questions. But as early as 1958, Chrysostom Constantides, a hierarch of the Patriarchate of Constantinople, published an article titled "Artificial Insemination and Theology" [11], in which he used Western natural law arguments to oppose the practice.

In Greece, a multi-volume *Encyclopedia of Religion and Ethics* was published between 1962 and 1968 [41]. A few bioethics articles were included in this work. Abortion was treated by Paul Polymeropoulos in a general and thoroughly negative fashion ([41], Vol. 5, pp. 539-540), and from the perspective of Orthodox canon law by George D. Kachrimanes ([41], Vol. 5, pp. 540-541). An anonymous entry on euthanasia addresses the issue from many different perspectives, but was generally opposed to the practice and to organizations seeking to promote the legalization of euthanasia ([41], Vol. 5, pp. 1027-1028). Interestingly, under the heading "death", there is a lengthy treatment from the perspectives of law, the history of religions, systematic theology, and laography (folklore), but not ethics ([41], Vol. 6, pp. 106-109).

Demetrios Constantelos contributed significantly to focusing attention on medical questions from the perspective of Orthodox church life and theology. In his ground-breaking historical study *Byzantine Philanthropy and Social Welfare* [8], he showed the historical connection between Eastern Orthodoxy, including the clergy [7], and medicine, an approach which was to bear fruit later for Eastern Orthodox bioethics.

II. WORKS ADDRESSING GENERAL BIOETHICAL QUESTIONS, 1970-1988

In this period, more voices began to express a nuanced Orthodox theological approach to birth control. The older view was still emphasized by some writers. Evangelos Mantzouneas, a specialist in Eastern Orthodox canon law, published a pamphlet expressive of this view under the title *Against Neo-Malthusians* [40]. In an effort to base an anti-birth control position on scientific opinion, George Malamos in 1975 published a volume of over three hundred pages, drawing on historical, sociological, and medical, as well as religious views, under the title *Birth Control in the Light of Science*.

Yet some of the authors writing during this period addressed the birth control issue in more positive terms. John Meyendorff, in his 1979 book *Marriage: An Orthodox Perspective* [42], was the first to do so in English. Others expressing this position were Demetrios Constantelos in *Marriage, Sexuality and Celibacy: A Greek Orthodox Perspective* [9], and Nicon Patrinacos in *The Orthodox Church on Birth Control* [46].

However, two studies in particular, rooted in the theology of the interpersonal relations of the Holy Trinity, the creation of human beings in the image and likeness of God, and the application of these presuppositions to marriage, led to a sustained frontal attack on Papakostas' rejection of the use of birth control in Christian marriage. In 1974 Chrysostom Zapheiris developed a sustained argument that proposed the view that birth control should be supported by the Orthodox precisely on the grounds that it could contribute to the fulfillment of the purposes of marriage [53]. Three years later Alexander M. Stavropoulos published a study on the 1937 Encyclical of the Holy Synod of Greece, in which he not only showed conclusively that its author was Papakostas, but also uncovered the non-Eastern Orthodox sources of Papakostas' approach [49].

Other questions began to be discussed in Orthodox circles. In the United States Constantine Zozos published a paper on sterilization in a student publication in 1974, raising points of opposition to the practice from an Orthodox perspective [55]. Stanley S. Harakas addressed a number of issues of bioethical praxis in newspaper columns published in the *Orthodox Observer*, the newspaper of the Greek Orthodox Archdiocese, from 1970 to 1975 under the *nom de plume* "Exetastes". A volume of these articles was published in 1976 under the title *Contemporary Issues: Orthodox Christian Perspectives* [22]. This volume, revised and enlarged, was reissued under the author's name six years later, entitled *Contemporary Moral Issues Facing the Orthodox Christian* [26]. Among the "contemporary moral issues" discussed were several on topics of concern to bioethics: venereal disease ([26], Sec. 20), contraception ([26], Sec. 21), abortion ([26], Secs. 22, 36), embryo fertilization outside the womb ([26], Sec. 23), death and dying ([26], Secs. 43, 44), euthanasia ([26], Sec. 45) and donation of organs after death ([26], Sec. 46).

A short but widely distributed pamphlet entitled *An Orthodox View of Abortion* was published in 1977 by John Kowalczyk [39]. It was rooted in the Greek patristic tradition of the Orthodox Church, showing the prohibition of the practice throughout the centuries, but also expressing pastoral concerns regarding the issue, as well as an appeal for Orthodox activism regarding the political aspects of abortion.

In 1978 the *Encyclopedia of Bioethics*, under the editorship of Warren T. Reich of Georgetown University, was published [48]. Harakas authored the articles on "Eastern Orthodox Christianity" [23] and the Eastern Orthodox portion in the general article on population issues [24]. In the former, Orthodox theological foundations for bioethics were sketched, including Eastern Orthodox theological anthropology, divine energies and human self-determination, and body-spirit relationships. The normative portion of the article was divided between concerns for the protection of life and the transmission of life. Under these two headings the following topics were discussed: health care, rights of patients, human experimentation, abortion, organ transplants, drug addiction, mental health, aging, death, dying, euthanasia, sexuality, fertility control, population, artificial insemination, genetic counseling and screening. The article was subsequently published as a small pamphlet [25].

The issue of genetic engineering and its ethical implications was addressed in 1983 by Athenagoras Nicholas Zakopoulos in his booklet *Contemporary Biology and Ethics* [51]. In this work Zakopoulos describes briefly the discovery of DNA and the potential developments of genetic engineering, with only a modest sense of possible dangers, while strongly encouraging the process. He formulates many questions about genetic engineering, ostensibly to provoke discussion, but he does little to indicate the direction an Orthodox ethical approach would take. His conclusion consists of four major points. First, he welcomes in a spirit of gratitude the newly discovered genetic knowledge provided by biology. Second, he insists that science must be left free to pursue genetic research with the assumption that its results will be shared responsibly and publicly. Third, he calls for "a sincere dialogue between...theologians, biologists, physicians, attorneys, psychologists, sociologists and representatives of the public" regarding the implications of DNA research for the general well-being of humankind. His final point rejects, however, the concept that biology can give ultimate significance and meaning to human life. He points to values that are produced by spiritual sources and that need to be related to the endeavors of genetics ([51], pp. 26-29).

A general restatement of Orthodox positions on several bioethical questions was written by Harakas in 1984 in a volume designed by the Greek Orthodox Archdiocese in New York for use by the media as a ready reference on the Orthodox Church. Entitled "The Stand of the Orthodox Church on Controversial Issues", the review appeared in *A Companion to the Greek Orthodox Church* [27].

The issue of euthanasia was addressed in 1985 by Thomas Hopko as a spokesman for the Russian Orthodox Church (i.e., Orthodox Church in America) ([38], pp. 55-57) and by Harakas on behalf of the Greek Orthodox Church ([38], pp. 45-56) in *Euthanasia and Religion* [38]. Both expressed the Orthodox Church's rejection of euthanasia. The latter rejected active euthanasia ethically, as a form of suicide, but held that "...the Orthodox Church would sanction the removal of the life-support system" in the case of a patient whose brain activity was irreversibly flat. "The reason for this is that [the patient] is already dead. The machines are simply keeping the dead body 'functioning'. ...Our church simply would not want to see the inordinate continuation of biological function after the person is dead" ([38], p. 45). Harakas provides guidelines for the Orthodox in facing issues regarding euthanasia, rejecting active euthanasia but developing tradition-based nuanced positions in the light of recent medical technological developments.

Evangelos K. Mantzouneas published his study *Organ Donations in the Orthodox Church of Greece* in 1985. In this 76-page work he discusses at length organ donation as a medical phenomenon, its place in the legal tradition, and its theological and canonical implications. After examining the issue from an ethical perspective, he concludes, "organ donations do not violate the canons of Christian ethics", but "every form of payment is prohibited". While focusing on duplicated organs, Mantzouneas also approves heart transplants, "because the heart as an organ for the circulation of blood for the body is a material thing. The heart about which Scripture speaks is a purely spiritual power" ([40], pp. 48-49). Mantzouneas notes that organ donations without financial reimbursement have been approved by Greek civil law and by the Holy Synod of the Church of Greece. He also documents that the present Archbishop of Athens, Serapheim, in order "to put words into practice", stated in his New Year Encyclical of 1985 that he has willed his kidneys and eyes for donation upon his death.

In 1986 an important step was taken in the United States for addressing the numerous questions of bioethics by the Orthodox Church with the organization of the "Orthodox Christian Association of Medicine, Psychology and Religion", under the guiding direction of Dr. John Chirban of Hellenic College, Brookline, Massachusetts. OCAMPR seeks to bring together physicians, psychologists, and persons with Eastern Orthodox religious identification in order to address concerns of therapeutic character in a wholistic manner. The Society publishes a bulletin titled *Synergia*, a Greek word meaning "cooperation", thus indicating its interdisciplinary approach. It holds regional meetings and an annual National Conference and is in the process of organizing a research center [6].

During 1986 Stanley Harakas published an historical study on the relationship of Orthodox Christianity to medicine [28], a paper on "Orthodox Christianity and Bioethics" [29], and a study on Orthodox perspectives regarding thalassemia, an illness with a high rate of incidence among Eastern Orthodox people in the Mediterranean area [30]. The first of these was an effort to trace the wholistic approach of Eastern Orthodox Christianity through the centuries, in its Greek and Slavic traditions, as it addressed the issues of health and well-being from both a medical and spiritual perspective. Characteristically, the Orthodox Church's approach is sacramental in both the larger and the more specific liturgical senses of the term. Both medicine and spiritual healing (primarily liturgical) coexist harmoniously in the tradition.

Striking evidence of this cooperative stance is seen in the now fully documented fact crediting the Orthodox Church tradition with the invention of the hospital as it is understood today – an institution organized for therapeutic purposes, using medical procedures in a systematic and organized way for healing persons with diseases. Conclusively proving this thesis is Timothy Miller's *The Birth of the Hospital in the Byzantine Empire* [43].

Athenagoras Zakopoulos, now a bishop of the Church of Greece, published a short study entitled *Euthanasia Today and the Position of the Church* in 1987. The author countered his perception of euthanasia as a form of murder with a strong statement calling for a caring and empathetic response toward those who suffer [52].

III. ORTHODOX BIOETHICS BIBLIOGRAPHY: 1988-90

The two-year period 1988-1990 saw the publication of some general approaches to the relationship of the Orthodox Church and its teaching to the broad realm of health and well-being. In 1988 Demetrios Constantelos published a general historical article titled "The Interface of Medicine and Religion in the Greek and the Christian Greek Orthodox Tradition" [10]. George Papademetriou made a contribution to the explication of the wholistic approach of Eastern Orthodoxy with his study on "The Human Body According to Saint Gregory Palamas" [44].

As will be seen below, given the essential sacramental approach of Eastern Orthodoxy, many bioethical issues are addressed from the perspective not of biology or abstract principle, but of the meaning of marriage within the sacramental understanding of the Orthodox faith. An excellent discussion of the Orthodox approach to marriage in broad ethical terms is included in Vigen Guroian's *Incarnate Love: Essays in Orthodox Ethics* [21], published in 1988. Guroian uses marriage as a vehicle to explicate the ethical tradition of Eastern Orthodoxy. While not bioethical in content, the volume contributes to an understanding of one of the trends in Eastern Orthodox thought which influences bioethical approaches.

Included among the general approaches to bioethical concerns was the publication in 1990 of Stanley S. Harakas' *Health and Medicine in the Eastern Orthodox Tradition* [34], a volume in the Park Ridge Center's series on the relations of various religious traditions to issues of health and medicine. Some attention was given to bioethical questions in this study, though the majority of the issues addressed explicate the wholistic approach of Eastern Orthodox Christianity in reference to a broad range of health and medical concerns. He also published articles on the themes of the therapeutic character of Eastern Orthodox understandings of salvation [32], on "rational medicine" in the Eastern Orthodox tradition [33], and on the "integrity of creation" and ethics [31].

Some articles also appeared during this period on the following selected bioethical topics. Other topics remained untouched by the Orthodox, who are now just beginning to grapple with these issues in sustained and careful treatments, as opposed to popular and exclusively pastoral writings.

1. *New Reproductive Techniques*

The range of reproductive techniques now being made available provoked discussions on several levels during the 1988-1990 period. Illustrative of the more popular, yet authentically Orthodox, treatments was an article in a diocesan publication by the Metropolitan of Peireus of Greece, Kallinikos, in March of 1990, "Ethical Problems Regarding Artificial Insemination" [36]. The major problem was not identified as the technical nature of the new procedures, as might be expected in a natural law pattern of thought.

Implicit, though not articulated, was an acceptance of medical intervention, presumably including artificial insemination by the husband (AIH), in assisting a married couple to conceive a child. The thrust of the article consists of an affirmation of the sanctity of the marriage and its integrity, composed of the total psychosomatic unity of its two partners and limited to them. "The union of the two in marriage is a great and holy mystery" (read "sacrament" for the term mystery). "A third party has no place in this relationship" ([36], p. 1).

Any form of reproductive technique, consequently, that violates the integrity and sanctity of the personal and psychosomatic relationship of the husband and wife is perceived as ethically inappropriate. Whatever process is used that causes the sperm of the husband or the ovum of the wife to be united with the sperm or ovum of a third person, through whatever technical means, or the use of the womb of another woman for the purpose of nurturing the fetus and bringing the child to term, creates a situation in which "we no longer have the creation of a child of a married couple, but the creation of a child which is the result of a decorous form of adultery (*euschemos moicheia*)", which is judged as "an ethically improper union with a third person" ([36], p. 1).

In an article entitled "Bio-Medical Technology: Of the Kingdom or of the Cosmos?" [4], John Breck deals with abortion, "anomalous forms of human procreation", and genetic engineering. In the section of his paper on "anomalous forms of human procreation", he discusses artificial insemination, in vitro fertilization, and surrogate motherhood (which he prefers to call surrogate gestation). Responding to Roman Catholic natural law approaches to the issue, Breck asks these questions for the Orthodox: "Do modern procreative technologies represent an unmitigated evil? Or may our faithful resort to them when conception does not occur through normal intercourse?" ([4], p. 11).

In his treatment of each of the three topics in his paper, Breck sees two important sides to the issues from his Orthodox perspective. The first, exemplified in the previous article, addresses the wholeness of the situation with its spiritual, personal, and somatic integrity, but Breck also wants serious consideration of the spiritual and personal issues of individual persons and their well-being, the *pastoral* dimension.

Thus, on the one hand he articulates a firm ethical evaluation of new reproductive technologies: "An unbridgeable ethical gulf separates artificial insemination that uses the sperm of the husband (AIH or 'homologous insemination') from that which uses sperm from a third-party donor (AID or 'heterologous insemination')". He ventures the judgment that "...in vitro fertilization can be considered morally acceptable only if it does not create 'extra embryos' that are destroyed, frozen for future use, or used for medical or commercial experimentation".

In reference to reproductive techniques that use, in varied forms, women as "host wombs", the judgment is similar: "As for surrogate mothering, the fact that a third party is necessarily introduced into the procreative act makes it totally unacceptable from an Orthodox point of view" ([4], p. 12).

Nevertheless, "given these limitations, it *may* be appropriate for a couple to resort to certain procedures such as artificial insemination by the husband

(AIH) and even in vitro fertilization, in order to conceive children of their own genetic heritage....To determine whether such procedures are appropriate or not in any given situation, however, requires a deep spiritual discernment that comes only through patience and prayer" ([4], p. 13). Breck is highly conscious of the dangers and exploitation inherent in these procedures and warns against them.

In a similar fashion, Breck deals with genetic engineering techniques, aware of some benefits, but even more aware of "the danger inherent in such experimentation", which he sees as two-fold: the "violation of the dignity and integrity of created life-forms, and interference in the evolutionary process of natural selection" ([4], p. 17). Thus, there is a cautious support of genetic engineering for therapeutic purposes, but not for eugenic or innovational techniques, especially as they refer to human beings. He expresses serious concern about the allocation of the resources for this therapy, and about the issues of control connected with it, questions regarding genetic screening and responsibility for errors and "deleterious effects" of the experimentation. He suggests, finally, limitations on the procedures, including a reversal of the right to obtain patents on new life forms. ([4], p. 22).

2. *Abortion*

By far the most frequently and most thoroughly discussed issue of bioethics among the Orthodox is abortion. An example of this concern in the popular religious press is seen in a series of articles in the newspaper of the Greek Orthodox Archdiocese, the *Orthodox Observer*, during 1990. The year began with the second part of an article begun in the last issue of 1989, written by Jim Golding, the editor, under the title "Abortion: 25,000,000 Dead...And Counting" [19]. The second part, published in January 1990, carried the title "Abortion From an Orthodox Christian Perspective" [20]. It was a clear condemnation of the practice on the basis of Eastern Orthodox traditional argumentation. In March, Edward Pehanich, the leader of Orthodox Christians For Life, built on the anti-abortion concerns with a two-part article "Infanticide: A Growing Threat" [47]. The Dean of the Greek Orthodox Cathedral in Boston, Athanasios Demos, provided a traditional counterpoint in an article "A Pastoral Approach to Abortion" [14] in November, calling for an understanding of the difference between a clear *ethical condemnation* of abortion, and the need for pastoral presence and support for parishioners caught in the problem. It was a classic Orthodox appeal to the principle of *oikonomia*. As noted above, John Breck also dealt with the issue of abortion in his 1988 article on biomedical technology [4]. It expressed traditional rejection of the practice with an appeal for pastoral concern in dealing with specific situations and cases.

The most extensive scholarly and analytical writing in the field of bioethics among the Orthodox during this period was a treatise published over the three-year period 1987-1989 in the Greek theological journal *Theologia*. Written by Chrysostom-Gerasimos Zapheires, the Metropolitan of Peristeri, it is a wide-ranging article, with 481 footnotes, drawing on extensive scriptural, patristic, legal, medical, theological, and other scholarly studies of international scope. It was published under the general title "Abortion and the Orthodox Church: Theses and Antitheses" [54].

The occasion for the writing of the article was the ultimately successful effort to legalize abortions in Greece. The introduction located the issue on moral grounds in relation to the existing laws regarding abortion and the efforts to change them. It described both popular and ecclesial reactions to these efforts, and gave critical commentary on the parliamentary bill then under consideration, in large part on the basis of the Church's status as the constitutionally recognized official religion of the Greek state.

The first major part of the study was a theological analysis of the topic, beginning in Chapter One with broad theological affirmations, leading to the affirmation of life as a supreme value. Life is characterized as a gift of the love of God and human beings as "co-creators" of life with God. An important part of Zapheires' argument is the wholistic affirmation of the unity of the spiritual and the physical in human life, and the understanding that in the tradition of Eastern Orthodox theological reflection this wholeness occurs at conception. This affirmation is then completed by the assessment of the abortion question in the light of Eastern Orthodox emphasis upon interpersonal communion and relationship, which leads the author to reaffirm the immorality of abortion.

The nature and substance of the embryo is the theme of Chapter Two of the study, in which the author affirms the personhood and autonomous existence of the embryo, based on biological, theological, and medical grounds. The fetus, according to Zapheires, is similar to all humans who are in the process of realizing their human potential in the image and likeness of God. The embryo is a human being "in the process of becoming", as is every other human being. Zapheires follows this with a discussion of the exercise of the mother's freedom *vis-a-vis* the human reality of the fetus, denying the mother life-and-death decisionmaking rights, since the fetus is not a part of her body, but another human being, "a person in the process of becoming".

Chapter Three characterizes abortions as premeditated murder, prohibited by the Canons and ethos of the Orthodox Church from the earliest period of its existence. He criticizes the medical profession for its complicity in the evil of abortion, as a denial of its mission and healing purpose. He concludes this part of his paper with a powerful rejection of the idea of abortion in general, and of its particular articulation in the bill then under consideration.

Part Two, "Basic Conditions for the Rejection of the Parliamentary Bill", begins in Chapter One with an assessment of the evil consequences of the legalization of abortion. Among these Zapheires discusses the violation of ethical criteria, issues of conscience, psychological problems, the encouragement of irresponsibility, the opposition to the creative work of God, the encouragement by means of the law of abortions for convenience, the issue of Greece's demographic problem (reduction of the population), potential medical problems for the woman having an abortion, the failure of legal restrictions to restrict the overall number of abortions, the use of abortion as a substitute for birth-control, and its encouragement of public immorality in general.

Chapter Two of this part of the study deals with the then existing penal code in Greece in reference to abortions, introducing the code provisions for self-defense into the abortion debate.

Part Three of the study discusses the supporters of abortion and their arguments. Regarding the first, in Chapter One Zapheires highlights feminists and those who claim oppression of the minority by the majority. Chapter Two discusses some of the reasons presented for the legalization of abortions in Greece. Here he addresses extensively the issues of the separation of Church and State in Greece, the relatively high incidence of abortions, and similar arguments.

In Part Four, Zapheires argues for measures which should be taken by both State and Church in a responsible manner in the face of the abortion question, calling for a realistic and honest approach by both agencies in a cooperative fashion. For example, he demands attention to the social injustices which often lead the poor to elect abortion as an answer to their inability to care for more children. In Chapter Two he discusses specific remedies, including support for large families by both State and Church through land grants and other such measures. He also argues for a systematic approach to the "unwanted child" and measures to strengthen adoption as a partial answer.

In a careful yet courageous section of this chapter, the author addresses the question of his Church's condemnation of birth control methods and counsels in a pastoral manner for its acceptance *kat' oikonomian* for the pastoral well-being of the people in the face of the greater evil of abortion. Placing the issue in the larger framework of marriage, the personal relations of spouses, and the sexual and the spiritual aspects of marriage, he argues that the ethical aspects of birth control methods are essentially the same as the periodic avoidance of spousal sexual relations taught by St. Paul. He presents many of the arguments of his 1974 article [53] to promote the use of birth control as a means to reduce the temptation to resort to abortion. The tone of his discussion of sexuality in marriage is pastoral and cast as an expression of love for the flock and its salvation. At the heart of his argument is a sacramental, interpersonal understanding of marriage.

The author further rejects the oft-repeated argument that only the woman has the personal right to decide about an abortion as a subjective, individualistic approach totally foreign to Christian perspectives. All these arguments are developed extensively to support family planning as an alternative to abortion.

The fifth and final part of the study challenges the Church to assume a leadership role in the nation and to address positively the issues raised by the legalization of abortion.

3. *Withdrawing/Withholding Care From Dying Patients*

On the issue of withdrawing or withholding care from dying patients, a brief "Statement by Stanley S. Harakas" was included as a feature article on "A Time to Die: The Cases of Nancy Cruzan and Janet Adkins" during 1990 [35], in which the author addressed the logic attending these cases and indicated the inadequacy of building moral cases on subjective choice.

An important treatment of the issue, originally presented at a national conference of the Orthodox Christian Association of Medicine, Psychology and Religion (OCAMPR) in 1989, was published that year by John Breck under the title "Selective Nontreatment of the Terminally Ill: An Orthodox Moral Perspective" [5]. Breck begins by examining three lines of ethical reasoning on the issue: (a) "vitalism", which "holds that biological life should be sustained at all costs"; (b) various forms of "euthanasia" that actively seek to bring unwanted life to an end; and (c) "non-treatment", allowing death to come "naturally". While drawn to the last, he identifies many "vexing questions" that remain unresolved by a simple statement of that position, especially the setting of limits even to the alternative of "allowing to die".

In order to address the question more fully, Breck seeks to place the question within the framework of the Orthodox understanding of human nature and the purpose of human life, leading to a realistic discussion of suffering both as potentially redemptive and as absurd and dehumanizing. He rejects both a vitalistic concern with keeping bodily functions "alive" and a view which disposes of life under a euphemism of "death with dignity". He proposes that the ultimate criterion for an Orthodox approach is the spiritual welfare of the patient.

One of the consequences of this approach is the affirmation that "the sanctity of life" approach and the "quality of life" approach "must no longer be seen as mutually exclusive....This means that we need to take with utmost seriousness the Church's traditional prayers for the peaceful separation of soul and body. When a patient is irreversibly comatose (insofar as can be determined) or suffering from intractable pain despite medication, then ethically responsible medical care will *not* retain as a primary goal perpetuation of that state" ([5], p. 268). The goal is not to prolong dying but to allow for a conscious preparation for a "good dying", understood as an opportunity for repentance and reconciliation with God and neighbor.

Breck has ambivalent feelings about withdrawing artificial feeding, recognizing on the one hand that such feeding may unnecessarily prolong the dying process, but fearing that even in the comatose state its withdrawal may provoke unnecessary pain. In any case, he rejects the alternative of active euthanasia. Since his primary criterion is the welfare of the patient, he is open to withdrawal of nutrition, but insists that one should not generalize in any absolute way. The last part of the article lists six practical steps to be taken by the ecclesial community in facing such issues, including the encouragement of hospice programs, living wills "whose only function would be to prevent non-beneficial or 'extraordinary' procedures from being imposed", and active church support for "comprehensive medical insurance for all its citizens".

4. *Active Euthanasia*

As has been made clear repeatedly in the issues previously discussed, the Orthodox Church is consistently opposed to any form of active euthanasia.

In the religious and cultural monthly review *Aktines* [3], during 1989 two articles were published by physician Athanasios B. Abramides on the subject of euthanasia. The conclusions showed a strong rejection of any form of active euthanasia and a hesitant, cautious sensitivity to the possibility in certain circumstances for a passive, "allowing to die" approach. The larger portion of both articles consisted of reports of attitudes on the topic. Of interest was the section on the "types of euthanasia", listing ten different gradations of euthanasia: active, passive, willed, unwilled, eugenic, social, non-resuscitative, psychological, living-will type, and narcotic.

5. *Organ Donation and Transplantation*

Little was written in this period on organ donations and transplants in Orthodox sources. In his general work on health and medicine, Stanley Harakas devoted a few pages to the topic. He reaffirmed the general Orthodox approval of the donation of organs after death and suggested that while redundant organs from living donors may be donated, "the Orthodox would hold that there is no moral law *demanding* organ donations, that the giving of an organ should be a free act of love". He raises the question of the donation of bodies to medical schools, which he defends with the proviso that appropriate funeral and burial rites take place, and that the remains be kept together for final interment ([34], p. 158).

Holy Cross Greek Orthodox School of Theology
Brookline, Massachusetts, U.S.A.

BIBLIOGRAPHY

1. Androutsos, C.: 1932, *Systema Ethikes* (Greek, *System of Ethics*), Athens.
2. Antoniades, B.: 1927, *Engcheiridion Kata Christon Ethikes* (Greek, *Handbook of Ethics According to Christ*), Vol. 1, Consisting of the General Part, and Vol. 2, Consisting of the Special Part, "Fazilet" Press, Constantinople.

3. Abramides, A.B.: 1989, "Euthanasia" (Greek, "Euthanasia"), *Aktines* 52, No. 498, 70-76; No. 499, 97-101.
4. Breck, J.: 1988, "Bio-Medical Technology: Of the Kingdom or of the Cosmos?", *St. Vladimir's Theological Quarterly* 32:1, 5-26.
5. Breck, J.: 1989, "Selective Nontreatment of the Terminally Ill: An Orthodox Moral Perspective", *St. Vladimir's Theological Quarterly* 33:3, 261-273.
6. Chirban, J. (ed.): 1990, "Research Center Meeting Held", *Synergia* 4, 2.
7. Constantelos, D.J.: 1966, "Physician-Priests in the Medieval Greek Church", *Greek Orthodox Theological Review* 11, 141-153.
8. Constantelos, D.J.: 1968, *Byzantine Philanthropy and Social Welfare*, Rutgers University Press, New Brunswick.
9. Constantelos, D.J.: 1975, *Marriage, Sexuality and Celibacy: A Greek Orthodox Perspective*, Light and Life Publishing Co., Minneapolis.
10. Constantelos, D.J.: 1988, "The Interface of Medicine and Religion in the Greek and the Christian Greek Orthodox Tradition", *Greek Orthodox Theological Review* 33, 1-18.
11. Constantinides, C.: 1958, "Technike gonimopoiesis kai theology" (Greek, "Artificial Insemination and Theology"), *Orthodoxia* 33, 66-79; 34, 36-52.
12. Cotsonis, J.: 1964, "Fundamental Principles of Orthodox Morality", in A.J. Philippou (ed.), *The Orthodox Ethos: Studies in Orthodoxy*, Holywell Press, Oxford.
13. Demetropoulos, P.: 1970, "Orthodoxos Christianike Ethike" (Greek, "Orthodox Christian Ethics"), n.p., Athens.
14. Demos, A.: 1990, "A Pastoral Approach to Abortion", *Orthodox Observer* (November), 6.
15. Dionysiatou, G.: 1957, *Mathousianismos: To Englema tes Genoktonias* (Greek, *Malthusianism: The Crime of Genocide*), Holy Mountain Library, Volos, Greece.
16. Evdokimov, P.: 1962, *Sacrement de l'amour: Le Mystere conjugal a la lumiere de la tradition orthodoxe*, Editions de l'Epi, Paris. Subsequently published in Greek in 1967, as *To Mysterion tes Agapes (The Mystery of Love)*, and in English in 1985, as *The Sacrament of Love: The Nuptial Mystery in the Light of the Orthodox Tradition*, St. Vladimir's Seminary Press, Crestwood, New York.
17. Galanopoulos, M.: 1954, *Systema Hieras Exomologeseos* (Greek, *A Study of Penitential Practice*), Orthodox Source Books, Athens.
18. Galanopoulos, M.: 1968, *Teknogonia* (Greek, *Childbearing*), Apostolic Diakonia of the Church of Greece, Athens.
19. Golding, J.: 1989, "Abortion: 25,000,000 Dead...And Counting", *The Orthodox Observer*, December 20, p. 27.
20. Golding, J.: 1990, "Abortion From An Orthodox Christian Perspective", *The Orthodox Observer*, January 10, p. 13.
21. Guroian, V.: 1988, *Incarnate Love: Essays in Orthodox Ethics*, University

of Notre Dame Press, Notre Dame.
22. Harakas, S. (*nom de plume*, Exetastes): 1976, *Contemporary Issues: Orthodox Christian Perspectives*, Greek Orthodox Archdiocese Press, New York.
23. Harakas, S.: 1978a, "Eastern Orthodox", *Encyclopedia of Bioethics*, in [48], Vol. 1, pp. 347-356.
24. Harakas, S.: 1978b, "Population: Eastern Orthodox Perspective", in [48], Vol. 3, pp. 1251-1254.
25. Harakas, S.: 1980, *For the Health of Body and Soul: An Eastern Orthodox Introduction to Bioethics*, Holy Cross Orthodox Press, Brookline, Massachusetts.
26. Harakas, S.: 1982, *Contemporary Moral Issues Facing the Orthodox Christian* (expanded and revised edition of [22]), Light & Life Publishing Co., Minneapolis.
27. Harakas, S.: 1984, "The Stand of the Orthodox Church on Controversial Issues", in *A Companion to the Greek Orthodox Church*, Greek Orthodox Archdiocese, New York.
28. Harakas, S.: 1986a, "Health and Medicine in the Orthodox Church", in R.L. Numbers, and D.W. Amundsen (eds.), *Caring and Curing: Health and Medicine in Western Religious Traditions*, Macmillan Publishing Co., New York, pp. 146-172.
29. Harakas, S.: 1986b, "Orthodox Christianity and Bioethics", in N.M. Vaporis (ed.), *Orthodox Christians and Muslims*, Holy Cross Orthodox Press, Brookline, Massachusetts, pp. 155-185; and in *Greek Orthodox Theological Review* 31, 181-194.
30. Harakas, S.: 1986c, "Thalassemia: A Theological Perspective", in J.T. Chirban (ed.), *Thalassemia: An Interdisciplinary Approach*, University Press of America, Lanham, Maryland, pp. 79-94.
31. Harakas, S.: 1988, "The Integrity of Creation and Ethics", *St. Vladimir's Theological Quarterly* 32, 27-42.
32. Harakas, S.: 1989, "Health and Medicine in the Eastern Orthodox Tradition: Salvation as the Context of Healing", *Greek Orthodox Theological Review* 34, 219-235.
33. Harakas, S.: 1989, "Rational Medicine and the Eastern Orthodox Tradition" (with Greek summary), in *Anaphora eis ten Mnemen Metropolitou Sardeon Maximou* (Greek, *Festschrift in Memory of Metropolitan Maximos of Sardis*), Holy Metropolis of Switzerland/Atef Danial Institute for Christian Unity, Chambesey, Geneva, Vol. 5, pp. 245-260.
34. Harakas, S.: 1990, *Health and Medicine in the Eastern Orthodox Tradition*, Crossroad Publishing Co., New York.
35. Harakas, S.: 1990, "Statement by Stanley S. Harakas", in "A Time to Die: The Cases of Nancy Cruzan and Janet Adkins", *Bulletin of the Park Ridge Center*, September, p. 24.
36. Kallinikos, Metropolitan of Peireus: 1990, "Problemata Ethikes Me ten

Techniken Gonimopoiesin" (Greek, "Ethical Problems Regarding Artificial Insemination"), *Peiraike Ekklesia* 104, 1 ff.
37. Karmiris, J.: 1967, *Christianike Ethike* (Greek, *Christian Ethics*), University of Athens, Athens.
38. Larue, G.A. (ed.): 1985, *Euthanasia and Religion: A Survey of the Attitudes of World Religions to the Right to Die*, The Hemlock Society, Los Angeles.
39. Kowalczyk, J.:, 1977, *An Orthodox View of Abortion*, n.p., Minneapolis.
40. Mantzouneas, E.: 1974, *Kata Neomalthousianon* (Greek, *Against Neomalthusians*), n.p., Athens.
41. Martinos, A. (ed.): 1962-67, *Threskeutike kai Ethike Engkyklopaideia* (Greek, *Religious and Ethical Encyclopedia*), A. Martinos Publishers, Athens.
42. Meyendorff, J.: 1970, 1975 (rev. ed.), 1984, *Marriage: An Orthodox Perspective*, St. Vladimir's Seminary Press, Crestwood, New York.
43. Miller, T.: 1985, *The Birth of the Hospital in the Byzantine Empire*, John Hopkins University Press, Baltimore.
44. Papademetriou, G.: 1989, "The Human Body According to Saint Gregory Palamas", *Greek Orthodox Theological Review* 34, 1-10.
45. Papakostas, S.: 1933, 1947, *To Zetema tes Technogonias: To Demographikon Problema apo Christianikes Apopses* (Greek, *The Question of the Procreation of Children: The Demographic Problem from a Christian Viewpoint*), Brotherhood of Theologians "Zoe", Athens.
46. Patrinacos, N.D.: 1975, *The Orthodox Church on Birth Control*, The Graphic Arts Press, Garwood, New Jersey.
47. Pehanich, E.: 1990, "Infanticide: A Growing Threat", *The Orthodox Observer*, March 7, p. 15; March 20, p. 23.
48. Reich, W.T.: 1978, *Encyclopedia of Bioethics*, The Free Press: Macmillan Publishing Co., New York.
49. Stavropoulos, A.M., 1977, *To Problema tes Technogonias kai He Engkyklios tes Ekklesias tes Ellados (1937): Symbole eis ten Poimantiken tes Teknogonias ex epopseos Orthodoxou* (Greek, *The Problem of the Procreation of Children and the Encyclical of the Church of Greece (1937): A Contribution to the Pastoral Concern for Child-Bearing from an Orthodox Perspective*), n.p., Athens.
50. Xristov, N., and I. Pantchovski: 1955, *Orthodox Christian Ethics* (Bulgarian), Sophia.
51. Zakopoulos, A.N.: 1983, *Synchrone Biologia kai Ethike: DNA – Genetike Mechanike: Prooptikes kai Anesychies* (Greek, *Contemporary Biology and Ethics: DNA – Genetic Engineering: Potentials and Concerns*), n.p., Athens.
52. Zakopoulos, A.N.: 1987, *He Euthanasia Semera kai he These tes Ekklesias* (Greek, *Euthanasia Today and the Position of the Church*), Apostolic Diakonia of the Church of Greece, Athens.
53. Zapheires, C.: 1974, "The Morality of Contraception: An Eastern Orthodox Opinion", *Journal of Ecumenical Studies* 11, 677-690.

54. Zapheires, C.: 1987, 1988, 1990, "Ai Ambloseis kai he Orthodoxos Ekklesia: Theseis kai Antitheseis" (Greek, "Abortion and the Orthodox Church: Theses and Antitheses") *Theologia* 58, 185-232, 385-424, 691-711; 59, 88-120, 246-277, 438-465, 660-711; 60, 7-72.
55. Zozos, C., 1974, "The Medical, Legal and Moral Aspects of Sterillzation", *Ekfrasis* (Spring), 33-54.

HASSAN HATHOUT

ISLAMIC CONCEPTS AND BIOETHICS

Islam is the third of the Abrahamic monotheistic religions, following Judaism and Christianity. Islam recognizes and respects Judaism and Christianity, in spite of certain doctrinal differences. The three religions share the same moral code, but Islam introduces *de novo* the framework of a comprehensive legal system to regulate and legislate for individual and communal human affairs. This total system is called the *Shari'a*, and is so flexible that the human (legal) mind can adapt it to variable times and places. The sources of the *Shari'a* are the *Quran* (to the Muslims, God's very words), the sayings and deeds of Prophet Mohammad, derivation of rulings by analogy, and finally, the consensus of Muslim scholars. The *Shari'a* is to serve the interests of the people, which are categorized as indispensable, necessary, and complementary. Each category has its set of rulings, maintaining the order of priority. The principal goals of the *Shari'a* are the preservation and protection of self (life, health, procreation, nourishment, etc.), of mind (relief from stress, psychological health, prohibition of drink and drugs, etc.), of ownership (sanctity of private ownership, legitimate pursuits of wealth and commercial laws, social function of capital, prohibition of stealth, fraud, and usury, etc.), and of honor (chastity, marriage and family laws, prohibition of fornication and adultery, social conduct, etc.). A set of juridically-recognized guidelines helps ethicists and legislators to derive rulings on issues not specified in the *Quran* or the authentic tradition of the Prophet, among which are: that necessities overrule prohibitions; the choice of the lesser of two harms; the priority of repelling harm over obtaining benefit; the priority of communal over individual welfare; and the rule of "wherever welfare goes, there goes the statute of God".

Until the near past, religious rulings on various issues have been the domain of the religious scholars (persons who in their education and studies specialize in religious sciences, although they are not clergy, since in Islam there is no church and no clergy). This worked well in older times, when scholars were encyclopedic personalities excelling in various branches of knowledge. This type of scholar became more and more rare as knowledge diversified into separate specialized fields that cannot be captured by the mind of the religious scholar on his own. Religious rulings in such fields can be achieved only by the combined efforts of scientists and religious scholars thinking together as one team. Efforts have been made to satisfy this need, above all by the Islamic Organization of Medical Sciences (IOMS) and the Fiqh Congress at Makka. The former is based in Kuwait, which at the time of this writing has been occupied and annexed by Iraq, inciting an international crisis the outcome of which remains to be seen.

I. CONTROL OF REPRODUCTION

A. *Contraception*

Islam permits contraception as long as it does not entail the radical separation of marriage from its procreative function. The evidence for the permissibility of contraception in Islam is documented by reports of companions of the Prophet who practiced coitus interruptus (the only contraceptive method known at the time), which the Prophet knew about but did not forbid. This practice, however, should have the mutual consent of the couple, according to the Prophet's ruling, "A man must not practice withdrawal with his wife without her consent". About 1000 years ago, the great scholar Al-Ghazali listed a full spectrum of valid indications for contraception, ranging from health reasons, where pregnancy would be a risk to the mother's health, through financial circumstances, to the mere preservation of the beauty of the woman's physique [9]. Although there are a few ultraconservatives, most people agree that, in principle, contraception is not prohibited. It must, however, be practiced by the choice of the couple and not imposed upon them by force or pressure. Any contraceptive method is acceptable, provided that it is not harmful and that it does not act by causing abortion.

1. *Reservations about Contraception*

The Middle East is a politically inflamed area, and the separation of politics from religion is alien to Islamic teaching and the Muslim mentality. Old colonialism and recent conflicts with the West have so sensitized people's feelings – political and religious – that for the masses anything coming from the West has to filter through a net of suspicious scrutiny. A number of reservations about contraception are therefore being voiced that no policy maker can afford to ignore.

It is often claimed that the disparity between resources and population growth is being overplayed. Statistics and projections seem less convincing when large amounts of surplus food and crops are dumped or burnt in advanced countries to maintain prices, while other nations suffer famine. The use of wheat exports as a strategic material for political leverage is cited.

It is also said that immense resources could yet be tapped if only a fraction of armament budgets were applied to doing so. The blame for the strained economy in some Muslim countries is often shifted from the population explosion to neo-colonialism in the form, for instance, of loans to Third World countries at interest rates so high that these countries can hardly service the loans, let alone repay them. The incitement of military threat or military conflict for the sake of highly profitable arms deals detracts from the West's prescription of curbing population growth as the only answer for limited resources. Indeed, some European countries that pioneered population limitation have begun to reverse their population policy and, because of a shortage of working hands, now to give incentives for larger families.

A new coinage in political jargon is the term "demographic warfare". This aims at manipulating population structure so as to reduce or eliminate politically or otherwise unwanted elements or, over the years, to boost a minority to become a majority, or to depress a majority to become a minority. No amount of preaching about health hazards of high parity or about the population time-bomb can have much effect under such circumstances. When people have lost everything but their sheer numbers, it is difficult to ask that they sacrifice those too.

Another remark relates to the relative safety of contraceptive drugs. Constant changes of formula and conflicting press coverage are universal phenomena, but when a preparation is prohibited in the country of manufacture yet is allowed to be exported for use in other countries, the situation becomes embarrassing.

2. *Breast-feeding as Natural Contraception*

It is surprising that breast-feeding has not been given due attention, since there are Islamic grounds in its favor. Breast-feeding causes the augmented release of a hormone, prolactin, which prevents ovulation. The *Quran* mentions breast-feeding and decrees that its full natural span is two years.

Breast-feeding is more than a nutritional process. It is also a "value" and a special bond, so much so that a woman other than the mother who breast-feeds an infant acquires a special relation in Islamic law, which is called "suckling parenthood", and she is the child's "mother in lactation". To accentuate its value, lactation motherhood is given the status of natural motherhood in legal rulings on marriage; the effect is that such a mother's natural children are considered siblings of the nursed child, who therefore may not marry any of them.

Although breast-feeding is not a sufficiently reliable contraceptive in individual cases because of the wide range of temporal variation in the resumption of ovulation, its total influence, if adopted by a society, would be great. It is estimated that, if breast-feeding became the normal way of feeding babies during their first two years of life, the drop in reproductive rate in a community would exceed that produced by all pharmaceutically produced contraceptive methods put together. Also, breast-feeding has well-known nutritional, health, psychological, and economic benefits, and for the policymaker, it is the contraceptive that will never arouse animosity in a religious community or be associated with colonialist or Western ill intentions. Family planning budgets and other resources might be best expended on a campaign to promote breast-feeding, recruiting religious leaders for what would qualify as a religious cause. For working mothers, radical departures from existing practice might entail changes in working hours or nursery facilities in the work place. These would be costly, but the benefits, in lower infant morbidity and mortality, as well as in population control, would compensate for the cost, not to mention positive effects in human happiness and better quality of life.

3. *The Intrauterine Contraceptive Device*

If the intrauterine device were an abortifacient rather than a contraceptive, it would be rejected. It was at first thought to act by preventing implantation of a fertilized ovum in the uterus. However, newer generations of the device contain a copper wire that releases spermicidal copper ions, or progesterone, which thickens cervical mucus, rendering it impenetrable to spermatozoa. A recent WHO Scientific Group concluded from the available evidence that the device impedes fertilization, and that it is therefore not an abortifacient [10]. It should therefore be acceptable and should play a principal role in the contraceptive armamentarium.

B. *Abortion*

Islam views abortion very differently from contraception, for abortion conflicts with the sanctity of human life, which is upheld by Divine Law. Commenting on the story of Cain and Abel (the two sons of Adam), the *Quran* (5:32) states:

> On that account We decreed upon the Children of Israel that whoever kills a soul for other [reasons] than manslaughter or corruption in the land, it shall be as if he killed all mankind, and whosoever saves one, it shall be as if he saved all mankind.

Human life is a value in itself, and should not be taken except upon the indications singled out and specified by the law (none of which falls within the domain of medicine).

The question that naturally arises is whether the term "human life" includes the intrauterine phase. According to the teachings of Islamic jurisprudence on the rights of the fetus, it does. Islam accords the fetus the status of "incomplete *zimma*". *Zimma* is the legal status that qualifies a person to have rights and to owe duties. It is incomplete in the case of the fetus, since it entails rights but not duties. Some of these rights are:

- If a husband dies while his wife is pregnant, the rules of inheritance recognize the fetus as an inheritor if born alive. The share of the unborn is set aside, and other inheritors upon receipt of their shares of the legacy give a documented pledge that if more than one baby is born, they will reimburse the share of all.
- If a fetus is miscarried at any stage of pregnancy and shows signs of life, such as a sneeze or a cough or movement, and then it dies, this fetus has the right to inherit anything it was legally entitled to inherit from anyone

who died after the beginning of the pregnancy; and, after it dies, what it has inherited is inherited in turn by its legal heirs.

- If a woman commits a crime punishable by death and is proven pregnant, then the execution of the punishment shall be postponed until she gives birth and nurses her baby until it is weaned. This applies irrespective of the duration of the pregnancy, however early, and implies the right of the fetus to life from its beginning. It applies even if the pregnancy is illegitimate, which shows that the fetus conceived out of wedlock also has the right to life. All sects and juridical schools unanimously uphold this ruling.

There is also a money penalty for abortion, even if it is inadvertent. This penalty, called the *ghorra*, equals one-twentieth of the penalty paid for inadvertent manslaughter. If aggression or willful action leads to abortion, the court will impose suitable punishment, on top of the financial penalty to be paid to the legal heirs of the aborted fetus.

The question of the beginning of life has been discussed since early times. Some ten centuries ago, Al-Ghazali rightly described a phase of imperceptible life before the fetus has grown enough to move inside the mother's womb [9]. He clarified the difference between contraception and abortion, in that the latter is an assault on an already existing life. "The first grade of existence," he wrote, "occurs when male matter falls into the womb and fuses with the female matter and gets ready to receive life". To destroy this is a crime, and the crime grows more and more serious as this matter passes from one phase to the other. The crime is more heinous after ensoulment, and is worst after the baby is born alive, as in the case of the pre-Islamic (*hiliya*) Arabs' practice of killing their children or burying their female neonates.

In older times, some juridical writers thought that until the mother felt quickening, the fetus had been lifeless flesh, and to them quickening meant "the instillation of life"; this usually occurred when the pregnancy reached four months. This also coincided with the date of "ensoulment", or the instillation of the spirit into the fetus by an angel. Other jurists thought that life began at 42 days, as the fetus took human shape. Modern science teaches that life antedates both events, and on scientific grounds, the stage of an individual's life that can be called its beginning should satisfy all the following criteria:

- it is a clear and well-defined event that can be pinpointed as the beginning of life;
- it exhibits the cardinal feature of the beginning of life, viz. growth;
- if this growth is not interrupted, it naturally leads to the subsequent stages of life as we know them: fetus, neonate, child, adolescent, adult, old person;
- it contains the genetic pattern that is characteristic of the human race at large, and also of a unique individual, of whom no other individual is a perfect copy, from eternity and for eternity; and,
- it is not preceded by any other phase which combines all the preceding characteristics.

Applying these criteria, life begins with the fusion of a spermatozoon with an ovum to form the fertilized ovum or zygote endowed with 46 chromosomes, half maternal and half paternal. Neither spermatozoon nor unfertilized ovum fulfills the criteria, although each is alive. Subsequent stages do not qualify because they are preceded by the zygote, which fulfills all the criteria.

The battle which rages in many parts of the world, especially in the United States, between the pro-life and the pro-choice movements is irrelevant to Islamic ideology. A woman does not create the fetus, but receives it; the fetus is entrusted to her. A fetus is not part of its mother's body and may indeed have a different blood group, rhesus, HLA, and genetic and biological attributes; and it cannot be considered as a tumor or a lump. A woman's freedom with regard to her body cannot entail the freedom to kill another human being, merely because it is inside her (usually at her own wish).

1. *When Is Abortion Permissible?*

Congresses attended by religious and medical scholars have concluded that human life is to be respected and protected from its beginning to its natural end, including its intrauterine phase (IPPF Conference on Islam and Family Planning, Rabat, Morocco 1970; Islamic Organization for Medical Sciences (IOMS) Symposium on Islam and Reproduction, Kuwait 1982). Only when the mother has a disease that makes the continuation of pregnancy a threat to her life is abortion permitted. In this case, the mother juridically is considered the root, the fetus the offshoot, and the offshoot may be sacrificed to save the root.

2. *Selective Abortion*

Abortion is unlawful if carried out because the fetus is of an unwanted sex (invariably the female). Besides the taking of innocent life, this act constitutes sex discrimination, which Islam condemns. It was known before Islam that some fathers would bury their female newborn alive, a crime the *Quran* (16:58-59) rebuked by saying:

> And when news is brought to one of them of the birth of a female child, his face darkens and he is filled with inward grief. With shame does he hide himself from his people because of the bad news about the baby: should he retain it and endure the contempt or bury it under the dust? Ah, how evil was their verdict.

Another type of selective abortion is often considered when a woman is carrying several fetuses (e.g., quintuplets), usually after the use of an ovulation drug for infertility. Because of the high risk of miscarriage, some obstetricians kill two or three of the fetuses by injecting a drug into their amniotic sacs, to make it more likely that the others will continue to maturity. According to many scholars, this is unacceptable because it involves the willful destruction of innocent life to ward off a danger that may never materialize, and other measures (such as bed rest or cervical cerclage) should be tried instead.

C. *Sterilization*

Sterilization usually entails the surgical interruption of the female or the male conduits. This form of surgery was not known at the beginning of Islam, and therefore there is no direct ruling on it, but the Islamic position is easily inferred. The Islamic version of human rights, referred to in juridical language as the "Goals of Jurisprudence", comprises the preservation of "self, religion, mind, property and procreation". In the scale drawn by Islamic law for compensation due as penalty for damages resulting from loss of life or of body members or functions, the penalty paid for causing loss of procreative function is equal to that for loss of life. This regard for the procreative function emphasizes the gravity of making a decision on sterilization.

When there is a clear medical indication, sterilization is not only permissible but may be necessary, such as when an incurable ailment makes further pregnancy a threat to the woman's life or health. Fear of transmission of genetic disease to the offspring can also be a valid indication, according to the principle of "repelling harm". When performed for non-medical reasons, however, the permissibility of sterilization becomes guarded, and every case should be carefully considered. For the young woman with low parity, it must be straight-forward rejection, but not so for the woman who is over 40, with a large number of children, whose fertility is practically spent, and who is suffering from the burdens and hazards of high parity.

The recent IOMS Symposium on Islam and Reproduction (Kuwait 1982) endorsed the following guidelines:

1. Sterilization should always be the outcome of voluntary enlightened consent of both spouses. No government policy should pressure people into sterilization, or exploit their poverty by monetary or other temptations.
2. It should be made clear that the decision on sterilization is final. No guarantee should be given that it can be reversed. Attempts to reverse the operation indicate an error in the primary choice of the method. The applicants should be warned of the possibility of widowhood, divorce, or loss of children. The limitations and expense of reversal surgery or in vitro fertilization in case of a change of heart should be honestly explained. Alternative reversible contraceptive measures should be suggested as a wiser option.
3. Wholesale sterilization should be banned, whether to curb population growth, to try new techniques, or to provide material for publication in academic papers.
4. Any doctor who is not convinced that sterilization is in the applicant's best interests has the right to decline.

In an Islamic hospital, the decision to sterilize is a serious one, to be taken by a senior member of the medical staff. With the expanding spectrum of safe and adequate contraception, sterilization tends to be discouraged, rather than encouraged.

II. TREATMENT OF INFERTILITY

It is legitimate for a married couple to seek a remedy for infertility if the marriage is barren, as long as the remedy does not conflict with the rules laid down by religion.

Scientific progress has led to a large expansion in methods and techniques that restore and improve fertility, but it is a feature of our scientific age to apply any method that can achieve the desired result without subjecting it to religious scrutiny; in other words, the end justifies the means. The bulk of Muslim people, however, still subscribe to the religious ruling, which was articulated with reference to some of the methods in current use in the 1982 IOMS Symposium on Islam and Reproduction.

A. *Artificial Insemination*

1. *Artificial Insemination by Husband's Semen (AIH)*

As long as it is necessary to effect pregnancy, this method is acceptable.

2. *Artificial Insemination by Donor's Semen (AID)*

The use of semen other than the husband's is not permissible, and if the husband is infertile beyond cure, then infertility should be accepted. Although AID is not adultery, it leads to procreation outside the specific duality of a marriage contract. There is a human right to be the legitimate and real child of one's own father, and to be the fruit of the bond of legitimate marriage. AID often may involve the lie of registering the baby as the child of a man who is not the natural father. Even if the husband and the wife are agreeable, such deception confiscates the rights of the newborn in legitimate ancestry. It unjustly absolves the real father from responsibility for his "flesh and blood", increases the risk of inadvertent incest in a community, and flouts juridical inheritance laws. The crux of the issue, of course, is the switch of the concept of "legitimacy" from God-made to man-made.

B. *In Vitro Fertilization (IVF)*

As long as IVF uses "embryo-replacement" rather than "embryo-transfer", it is acceptable to Islam as a treatment of infertility. It is legitimately performed between husband and wife while they are still husband and wife (not after widowhood or divorce). Intervention of a third party other than husband and wife and the bearers of their genetic material (sperm and ovum) would not be permitted. Reference has already been made to "alien semen", donor semen which is not the husband's, fertilizing a woman. The same applies to a donated "alien ovum", or to an "alien embryo" carried by a woman's uterus and not entirely made from her own and her legitimate husband's genetic material.

C. *Surrogate Motherhood*

One form of surrogate mother is the woman who is impregnated by the sperm of a man whose own wife cannot become pregnant. In this case, the resulting baby's genetic material will be partly hers and partly the man's. The other form occurs when the wife can produce an ovum but cannot carry the embryo through pregnancy. In vitro fertilization is carried out between husband and wife, and the resulting embryo is deposited in the uterus of a surrogate mother for the rest of the pregnancy. In either case, the child will be handed over to the husband and wife and registered as their own child, and the surrogate who carried and gave birth to the baby will legally give away all rights and claims, usually for an agreed fee.

Surrogate motherhood is absolutely unacceptable to Islam. There is even more to it than denying to the progeny the right to a legitimate root, or deceiving both children and the civil registries about real parentage and lines of ancestry, etc. For perhaps the first time in the history of mankind, a woman agrees and plans to become pregnant with a child, knowing that she will give the child away. It is worst of all if done for money, thus reducing motherhood from a "value" to a "price". If this becomes widespread, it is not difficult to comprehend its shattering general effect on the regard of children towards parents and the emotional ties between each generation and its predecessor. If social norms ever permit children to become commodities, manipulated with a "buy and sell" rule, values other than parenthood will dwindle, values that we much cherish, such as natural love and mutual compassion and care within the family unit.

III. COMMENT

Some conclusions of a combined panel of religious scholars and medical scientists on "Human Reproduction in Islam" [3] are worth mentioning:

Fetal Sex Selection: There was an agreement that the Islamic viewpoint is that fetal sex selection is unlawful when it is practiced at a national level, [w]hile some of the scholars participating in the seminar believe that on an individual basis there is...nothing legally wrong in the attempt to fulfill the wish of a married couple to have a boy or a girl through available medical means,...other scholars believe it is unlawful lest one sex might outnumber the other.

Test Tube Babies (and Surrogate Motherhood): The seminar concluded that such a practice is lawful only if it involves a married couple while the marriage is valid and sufficient, and meticulous care is taken to avoid confusion of lineage, although even here some members expressed reservations. It was also agreed that the practice is unlawful if it involves an alien party, whether in the form of semen, an ovum, an embryo, or a uterus.

Surgical Contraception (Sterilization): It is agreed that surgical contraception is lawful on the individual level in cases of necessity as determined by a trustworthy Muslim doctor and when other alternatives have been exhausted. On the level of Muslims at large, it is unlawful. The Seminar denounced turning sterilization into a general campaign, and warned against its exploitation in demographic wars that aim at turning Muslims into minorities in their own countries or in the world.

Abortion: Going over the views expressed by earlier Fiqh scholars, with the keen insight and sound judgement they demonstrate, and noting that they unanimously forbid abortion after the breathing in of the spirit, i.e., after the first four months of pregnancy, [but differ on the ruling before this,]...the seminar concluded that an embryo is a living organism from the moment of conception, and its life is to be respected in all its stages, especially after spirit is breathed in. Aggression against it in the form of abortion is unlawful except in cases of maximal necessity. Some participants, however, disagreed, and believe that abortion before the fortieth day, particularly when there is justification, is lawful.

In a 1989 symposium in Kuwait, jointly held by the IOMS and the Fiqh Congress of Makka, the fate of surplus fertilized ova from in vitro fertilization procedures was discussed. There was unanimity that the ideal situation was to improve the techniques of cryopreservation of "unfertilized" ova, and that on each attempt at fertilization, only the number of eggs to be lodged in the uterus would be exposed to the sperm. In this case, no surplus fertilized ova would be allowed to form. If no ova are fertilized following exposure to semen, the attempt is repeated on the following cycle. This would entail more time and money, but both have less value than human life.

Opinion was divided about what to do with surplus ova resulting from current techniques. Some participants thought that surplus ova have no sanctity until implanted in the uterus and therefore can be destroyed; some saw no objection to using them for scientific research until two weeks of age. Other participants considered these ova to be human beings from the earliest stages, and thought that allowing them to die naturally would be preferable to either destroying them or using them for research.

IV. DEFINITION OF DEATH

The definition of the moment of death was exhaustively discussed in the seminar "Human Life: Its Inception and Its End as Viewed by Islam", the proceedings of which have been published [7]. The implications bear not only on the feasibility of procuring vital organs for transplantation, but on other juridical issues such as determining the beginning of the waiting period a widow has to honor before remarrying, and the apportioning of legacy shares if two or more inheritors should die in succession. Medical scientists explained the concept of brain death, which initially evoked reservation among the religious scholars, who preferred the traditional diagnostic signs of death, finding it objectionable to consider as dead a person who still exhibits such signs of life (albeit by artificial assistance) as a beating heart, metabolism, nutrition, excretion, and hair growth. The deadlock was overcome when the medical side forwarded an analogy based on a traditional juridical ruling, viz., "the movement of the slain". It was an old ruling that if an aggressor stabs a victim, with the result that the latter's intestines are extruded from his abdomen, this was considered a fatal injury (which it invariably was in the era during which the ruling was made), and the movements made by the victim (in the throes of death) had the juridical technical name of "the movement of the slain". If a second aggressor then finished the victim off, the murder charge would still lie against the first aggressor, who dealt the fatal injury, while the second would be charged only with aggression against the cadaver. The medical side proposed that, although today the extrusion of bowels following a stab wound is not necessarily fatal, inasmuch as it can be successfully repaired by surgery, the condition equivalent to "the movement of the slain" is that of a person whose brain (including brain stem) is dead, even if the rest of the body is still being maintained artificially. Differential diagnosis and the question of diagnostic certainty then followed.

The final recommendations were as follows ([7], pp. 628-629):

- The seminar realized that in the majority of cases there is no difficulty in recognizing the occurrence of death through conventional signs or as the result of external medical observation which notes the absence of the signs that distinguish the living from the dead.
- The seminar reached the conclusion that there are a few cases that have particular importance, because there is an urgent need to diagnose them as cases of death, although the body still shows signs which have been always accepted as signs of life, whether naturally displayed by some organs of the body, or resulting from resuscitation equipment applied to the patient.
- The seminar discussed the signs of death listed by Fiqh reference works and discovered that in the absence of a *Quran* or Tradition text that explicitly defines death, opinions reflect the medical knowledge available at the time of writing. Because the diagnosis and the signs of death have always been a medical matter on the basis of which Fiqh scholars make legal rulings, participant physicians presented the current medical view concerning the occurrence of death.
- From the presentations of physicians, it was clear that the decisive factor for them in proclaiming the death of a patient is the lifelessness of the area of the brain that is responsible for vital body functions, which they express as death of the brain stem. A diagnosis of brain stem death has clear conditions, which rule out certain suspicious cases, and physicians are capable of coming up with diagnoses of brain stem death that are reasonably certain. Any other vital organ or function, such as the heart or respiration, may temporarily stop, but as long as the brain stem is alive, one should try to revive the patient. The lives of a number of patients are saved this way.

When, however, the brain stem itself is dead, there is no hope of reviving the patient, even if other systems of the body continue to move or function.

- On the basis of this presentation by the physicians, Fiqh scholars were inclined to the view that when it is ascertained that a human being has reached the stage of brain stem death, he is considered to have withdrawn from life, and certain rulings of the dead are applicable to him, in analogy – though with evident differences – with what Fiqh books say about an injured person who has reached the stage known as that of the "slain". As for the remaining rulings that concern the dead, the participant Fiqh scholars prefer their postponement until all major systems of the body come to a stop. It was recommended that an additional detailed study should be made to determine which rulings for the dead apply immediately and which should be delayed.
- On the basis of the above, it was agreed that when the death of the brain stem is certified by a report of a committee of specialists, it is lawful to remove the resuscitation equipment.

V. ORGAN DONATION AND TRANSPLANTATION

The Islamic Code of Medical Ethics endorsed by the First International Conference on Islamic Medicine reads as follows ([5], pp. 81-84):

The individual patient is the collective responsibility of society that has to ensure his health needs by any means, inflicting no harm on others. This comprises the donation of body fluids or organs such as blood transfusion to the bleeding or a kidney transplant to the patient with bilateral irreparable renal damage. This is another 'Fardh Kifaya', a duty that donors fulfill on behalf of society.

Organ donation shall never be the outcome of compulsion, family embarrassment, social or other pressure or exploitation of financial need.

Donation shall not entail the exposure of the donor to harm.

The medical profession bears the greatest portion of responsibility for laying down the laws, rules and regulations organizing organ donation during life or after death by a statement in the donor's will or the consent of his family; as well as the establishment of tissue and organ banks for tissues amenable to storage. Cooperation with similar banks abroad is to be established on the basis of reciprocal aid.

Umar ibnul-Khattab, second Caliph after the Prophet, decreed that if a person living in a locality died out of hunger, being unable of self-sustenance, then the community should pay his money ransom...as if they had killed him. The similitude to people dying because of lack of blood transfusion or a donated kidney is very close.

Two traditions of the Prophet seem to be quite relevant in this respect. The one is 'The faithful in their mutual love and compassion are like the body...if one member complains of an ailment, all other members will rally in response.' The other tradition says, 'The faithful to one another are like the blocks in a whole building; they fortify one another.'

God describes the faithful in the Quran: 'They give priority over themselves even though they are needy.' This is even a step further than donating a kidney, for the donor can dispense with one kidney and live normally with the other...as routinely ascertained medically prior to donation.

If the living are able to donate, then the dead are even more so, and no harm will afflict the cadaver if heart, kidneys, eyes or arteries are taken to be put to good use in a living person. This is indeed a charity...and directly fulfills God's words: 'And whosoever saves a human life it is as though he has saved all mankind.'

A word of caution, however, is necessary. Donation should be voluntary by free will...or the dictators will confiscate peoples' organs, thus violating two basic Islamic rights: the right of freedom and the right of ownership. In the society of the faithful, donation should be in generous supply and should be the fruit of faith and love of God and His subjects.

The code was subjected to further scrutiny in a symposium on "The Islamic Vision of Some Medical Practices" [7]. Papers were presented there on "Donation, Sale and Bequeathal of Human Organs" [6], "Legal Ruling on Sale or Donation of Human Organs" [10], "Responsibility of Doctors as Viewed by Jurists" [11], "Disposal of Human Organs" [1], "Sale of Human Organs" [13], and "Sale of Human Organs in the Balance of Legitimacy" [2]. An exhaustive discussion followed ([8], pp. 353-383).

Various juridical themes were clarified. The human body is divinely endowed with respect and honor, and is not to be violated whether alive or dead. We do not own our bodies, but we are entrusted with them to protect and maintain them and use them only in the ways permitted by the Creator. For the purpose of organ donation and transplantation, two juridical rules are brought into effect: the first is that "necessities overrule prohibitions", and the second is "the choice of the lesser of two evils". The living or the dead human body can be invaded (which is evil) to save the life of a patient (which is necessary), lest he die (which is the more evil alternative). It is imperative that the living donor should not be threatened by the donation, either in his life or in its quality (such as a living person donating both corneas).

Donation should be free and completely voluntary. Because of the honor accorded the human being by God, human organs should not be treated as commodities for sale. If there is dire need, and purchase of the organ is the only way it can be procured, then it is lawful for the needy buyer (by necessity), although sinful for the seller, unless the latter is extremely needy. A system of suitable reward may be devised, and the service is better entrusted to a governmental authority to avoid organ trafficking in an exploitative market that would certainly exclude the unwealthy. On the sale of organs, the recommendations of the symposium included:

- The best way to obtain organs is through people's compassion for each other leading to donation of organs from dead bodies, either by a will or by the approval of heirs, as well from dead people with no known relatives.
- In the opinion of the majority, it is also permissible to obtain organs by donation from living persons, provided that adequate controls are followed,

organs.

- Sale of organs is prohibited. However, if organs cannot be obtained by donation and have to be paid for, then payment is permissible by the approval of the majority of participants. It is a prohibition made permissible only under necessity. Some participants held to the opinion of prohibition.
- In all cases, and especially in critical ones, obtaining organs should not be left to the competition between rich and poor recipients; the state should establish an authority to control procurement and to avoid its perils, managing it according to a prescribed detailed law.

Other aspects of organ transplantation were further discussed in a joint meeting of the IOMS and the Fiqh Congress of Makka held in Kuwait, October 23-26, 1989, on "Islam and Current Medical Issues". The following is a summary of its recommendations:

- Transplantation of Brain and Nervous Cells: This was recently described as promising in treating some diseases. It is lawful if the source is the suprarenal gland medulla or an animal fetus, or a human fetus spontaneously miscarried when it dies naturally. It is unlawful to sacrifice a living or viable human fetus for the purpose. If attempts at tissue culturing nervous cells succeed, the problem would be solved. In lawful abortion (to save the mother's life), the fetus may be used. Creating fetuses and performing abortion for the purpose of transplantation is unlawful.
- The Anencephalic Fetus: As long as it lives it should not be used as a source of organs for transplantation. Artificial termination of its life is unlawful. It may be maintained by artificial resuscitation to keep its tissues healthy until its brain stem dies, and then it is legitimate to take its organ.
- Transplantation of Sex Glands: It is unlawful to transplant testes capable of producing spermatozoa, or ovaries capable of producing eggs, into another person, for this would lead to confusion of genealogy and the conception of babies by gametes that are not united by an authentic marriage. Sex glands that are infertile (do not produce gametes), but are hormonally productive, are not subject to this ban, but their use for hormone substitution is only hypothetical, and legitimate only insofar as it can be achieved medically without subjecting the patient to the phenomenon of graft rejection or to the hazardous therapy it calls for. Sex organs that do not lead to genetic transmission may be transplanted if there is a lawful necessity, as recommended by the majority against an opposing minority.

Islamic Center of Southern California
Los Angeles, California, U.S.A.

BIBLIOGRAPHY

1. Faidullah, M.: 1989, "Disposal of Human Organs", in [8], pp. 306-320.
2. Foutouh, M. abul-: 1989, "Sale of Human Organs in the Balance of Legitimacy", in [8], pp. 342-352.
3. Gindi, A.R. al- (ed.): 1989, *Human Reproduction in Islam*, trans. A.M.M.S. Asbahi, IOMS (Islamic Organization for Medical Sciences) and KFAS (Kuwaiti Foundation for the Advancement of Science), Kuwait.
4. Hathout, H.: 1989, "On Ethics and Human Values in Family Planning", in Z. Bankowski *et al.* (eds.), *Proceedings of XXII CIOMS Conference on Ethics and Human Values in Family Planning*, CIOMS, Geneva, pp. 222-237.
5. IOMS (Islamic Organization for Medical Sciences): 1981, *First International Conference on Islamic Medicine, Proceedings*, IOMS, Kuwait.
6. Mahdi, M. el-: 1989, "Donation, Sale and Bequeathal of Human Organs", in [8], pp. 280-286.
7. Mazkur, K. al-, *et al.* (eds.): 1989, *Human Life: Its Inception and Its End as Viewed by Islam*, trans. M.M.S. Asbahi, IOMS (Islamic Organization for Medical Sciences) and KFAS (Kuwaiti Foundation for the Advancement of Science), Kuwait.
8. Mazkur, K. al-, *et al.* (eds.): 1989, *Islamic Vision of Some Medical Practices*, IOMS (Islamic Organization for Medical Sciences) and KFAS (Kuwaiti Foundation for the Advancement of Science), Kuwait.
9. Shaltout, M.: 1966, *Islam: Creed and Law*, Darul Qalam, Cairo.
10. Tantawi, M.S.: 1989, "Legal Ruling on Sale or Donation of Human Organs", in [7], pp. 306-320.
11. Tantawi, M.S.: 1989, "Responsibility of Doctors as Viewed by Jurists", in [7], pp. 297-305.
12. World Health Organization: 1987, *Mechanism of Action, Safety and Efficacy of Intrauterine Devices: Report of a WHO Scientific Group* (WHO Technological Report Series, No. 753), World Health Organization, Geneva.
13. Yaseen, M.N.: 1989, "Sale of Human Organs", in [8], pp. 321-341.

PAUL NELSON

BIOETHICS IN THE LUTHERAN TRADITION

What do Lutherans say about issues in biomedical ethics? How do they approach such questions? Is there a distinctively Lutheran perspective on bioethics? This report addresses these questions with respect to the teachings and policies of the major Lutheran churches in the United States. Readers are cautioned against the assumption that positions taken by the churches necessarily represent a consensus, or even the majority view, among those approximately ten million Americans who call themselves Lutherans. One should realize, also, that these official teachings are expressions of the views of bureaucratic departments, committees, commissions, and ultimately church conventions, in which the influence, and even more the control, of confessionally Lutheran theology is greatly attenuated by other considerations. It is not uncommon to find prominent Lutheran theologians and ethicists in vigorous dissent from the "official" positions of their churches. While it is not possible to survey the writings of individual Lutherans within the *Yearbook*'s allotted pages (except when their work was published by the churches as study material), some of them do appear in the notes.

I. THEOLOGICAL ETHICS IN THE LUTHERAN TRADITION

"A living tradition", according to Alasdair MacIntyre, "is an historically extended, socially embodied argument, and an argument precisely in part about the goods which constitute the tradition" ([32], p. 222). Traditions are places of agreement and disagreement, continuity and change, and of conversation among people who, despite all their differences, share some common language and loyalty. The Lutheran theological tradition conforms to this description. Lutherans understand themselves to be members of the "one, holy, Catholic, and Apostolic church" identified in the Ecumenical Creeds of Christendom, and they recognize the Augsburg Confession of 1530 and the other documents collected in the *Book of Concord* (1580) as valid, authoritative interpretations of the faith of the church catholic. Although the confessional writings are generally consonant with the teachings of Martin Luther (indeed, he drafted some of them), the reformer's views are not themselves authoritative. Perhaps it is fortunate that Luther did not think of himself as the founder of Lutheranism, because some of his opinions and practices can only be repudiated. In any case, the church documents to be examined here do not try to warrant their judgments by citing Luther.

Nevertheless, in many respects the tradition continues to speak Luther's language. His cathartic insight that believers are "justified" before God "by grace through faith" in Jesus Christ, and not on account of their good "works", shapes the Lutheran conception of the moral life. Liberated from enervating preoccupation with achieving their own salvation through the performance of "works of law", Christians are empowered to love and serve others. This follows the biblical pattern established by St. Paul, wherein the Christian life is one of "faith working through love" (*Gal.* 5:6). The believer's newfound Christian liberty finds its proper exercise in spontaneous self-sacrificial service to the neighbor. Neighbors and their needs are encountered in the family, in the work place, and in the state; orders of creation are ordained and governed by God's law for the well-being of all. While it is the gospel which reconciles the Christian with God, the law continues its important work of restraining moral evil and convicting of sin, thereby throwing sinners upon the divine mercy.

In a society still oppressed by late medieval clericalism and rampant "works righteousness", Luther's teaching was a welcome corrective. However its historical influence is a mixed blessing in that (besides dividing the Western church) it introduced into the Lutheran theological tradition a marked ambivalence concerning the status of ethics as a theological discipline. Consider the following:

In opposition to all unevangelical ethics of principles, "blue laws", ideals or rules and regulations ...a Christian ethic based on the "divine indicative" of God's grace (rather than the "divine imperative" of God's law) preserves the freedom of the believer under the guidance of the Holy Spirit through the Bible, the church, and prayer, to discover anew in each concrete situation what the will of God permits or requires of him or her then and there ([25], p. 361).

This depiction of Luther's evangelical ethic by a respected Lutheran ethicist is representative of the tradition's aversion to any sort of moral reasoning in which can be detected the slightest whiff of casuistry. Precise moral reasoning is sometimes held suspect as an enterprise of human self-justification or "works righteousness". To "sin boldly" and pray for forgiveness (which you will need in any case) is often held to be the better course.

The Lutheran penchant for intuitionism and occasionalism is evident in the Danish theologian Knud Løgstrup's interpretation of Jesus' teaching concerning love of neighbor. In a chapter entitled "The Impossibility of Christian Ethics", Løgstrup states:

It is characteristic of everything that Jesus said – so far as it has been transmitted to us – of every story and parable, of every one of his answers in conversation or argument, of every concisely formulated utterance, that it is a proclamation of the demand which in itself is silent. This is the intangible in his proclamation, that which foredooms all efforts to systematize it. Jesus' proclamation contains no directions, no rules, no moralizing, no casuistry. It contains nothing which relieves us of responsibility by solving in advance the conflicts into which the demand places us. All of his words speak about the one demand, but not with as much as a single syllable does he break its silence ([28], p. 115).

I have quoted these lines at length because they nicely illustrate two of the principal Lutheran criticisms, namely, that the casuistical enterprise is foreign to the New Testament, and that it inevitably leads its practitioners into self-righteous complacency.

I shall put aside the first issue, since a proper assessment would require a larger excursion into biblical theology than is practical here. However, it may be observed that in itself it is not a compelling reason for a Lutheran to reject casuistry. Such appeals to Scripture in restricting either ecclesiastical practice or theological formulations are themselves at odds with both the teachings of the Lutheran Confessions and the practice of the 16th-century Lutheran reformers. In any case, it is the second criticism which has the greater currency and significance.

Like many Lutherans, Løgstrup suspects that any use of principles or rules is an attempt to avoid responsibility for the radical demands of an eschatological ethic. Were the divine will for persons to be given specific content or, as he says, codified into a "worldview", Christians could content themselves with an unreflective, external compliance "without any investment of [their] own humanity, imagination or insight" ([28], p. 116). When this happens,

> We pretend to possess a divinely attested knowledge of what is to be said and done in the given situation and of what our mutual relationships ought to be. We have thus turned God into the legal, moral and political reason for what we should say and do. We break the silence often in a very noisy and self-opinionated manner and with an unbearable and loquacious sense of superiority ([28], p. 117).

In contrast, it is Løgstrup's conviction that we can never be entirely sure that we have acted in the right manner. Moreover, he judges that this inescapable uncertainty plaguing all our reflection and action is somehow our own fault. (More confessional Lutherans are not reticent about attributing this to fallenness and sin.) Yet neither Løgstrup nor other Lutheran critics of natural law, such as Helmut Thielicke, have demonstrated that contemporary modest forms of casuistry ineluctably lead to the spiritually perilous corruptions against which they polemicize. At the very least, they have not shown that a chastened casuistry is more dangerous than the ills to which Christian intuitionist or existentialist ethics are susceptible.

Løgstrup's own proposal for understanding the Christian moral life is based upon his perception that "trust" is a fundamental part of human life and a "fact of creation" ([28]), pp. 9, 21). Since, as he is fond of saying, "we hold each other's lives in our hands", we must acknowledge the demand this places upon us to "take care" of the "neighbor's well-being". The tides of existentialism, occasionalism, and intuitionism run high in that Løgstrup is very wary of attempting to specify the content of this "caring". No doubt it is the Lutheran in Løgstrup that wants to believe that "a good tree bears good fruit". But this needs to be juxtaposed with the equally Lutheran conviction that few, if any, trees are unqualifiedly "good" in the requisite sense.

In the end, it seems that Løgstrup has a rather optimistic view of human nature. While he acknowledges the possibility of self-deception, he believes that it is actually "a product of morality":

Self-deception presupposes the conscience from which it flees....Conscience tears everything aside and points to one's guilt....This is the intuitive aspect of the conscience ([28], p. 162).

It is almost as if conscience is, for Løgstrup, a special "moral sense" or faculty which perceives what the ethical demand requires. Yet this is somewhat ironic in that it is the insight afforded by conscience to which natural law casuistry traditionally appeals for its foundation. Furthermore, it is precisely this sanguine appraisal of the uncorrupted conscience which Lutheran theologians like Thielicke emphatically contest. In fact, their rejection of casuistry rests in large part on their assertion that as a consequence of the Fall, conscience is no less corrupt than any other feature of our humanity. Were these Lutherans more consistent, they might recognize the dangers this conviction ought to signal for intuitionist, occasionalist ethics.

These Lutheran depictions of the moral life might prompt Roman Catholic, Reformed, and other observers to ask, "Can ethics be Lutheran?" The answer depends on what is taken to be normatively Lutheran. For example, if the Reformation principle of *sola fide* (faith alone), along with Luther's critique of 16th century Catholicism and his own moral vision, are thought sufficient to constitute for all time an adequate Christian ethic, then the answer is negative, as Løgstrup illustrates. However, if these Lutheran concerns are seen in their historical context as critical correctives applied to Catholic "substance" (which was Luther's understanding), then certainly it is possible, even necessary, to engage in careful moral reasoning without abandoning confessional Lutheranism. Since the confessions prudently refrain from mandating a particular method of moral reasoning, their corrective principles may be considered permanently valid cautions for any style of Christian ethics. Understood in this way, the Lutheran confessions allow ethicists considerable methodological latitude, so long as the *sola fide* is not violated. Even such putatively suspect ideas as natural law, which, it should not be forgotten, Luther himself employed, can contribute to a properly evangelical (gospel-centered) Lutheran ethic. In any case, Lutherans are obliged to exercise moral reason and conscience (by no means the exclusive possession of Christians) as they and their non-Christian neighbors seek to discern and pursue the common good.

II. THE LUTHERAN CHURCHES

In the United States today Lutherans are divided among two large and seven much smaller churches. The largest group, the Evangelical Lutheran Church in America (ELCA), has 5.3 million members in over eleven thousand congregations, and is affiliated with 270 hospitals, homes for the elderly, and other social service facilities for persons with special needs. The Lutheran Church-Missouri Synod (LCMS), more conservative than the ELCA on many theological and moral issues, has 2.6 million members in nearly six thousand congregations. Collectively, the smaller groups have about half a million members in 1,700 congregations.

The ELCA was formed in 1988 by the merger of the Lutheran Church in America (LCA), the American Lutheran Church (ALC), and a much smaller group of "liberals" who had left the LCMS a few years before in a controversy concerning the authority of scripture. As the ELCA begins to develop its own teaching on ethical issues, it has elected to receive the social statements of the ALC and LCA and encourage their use. For this reason the policies and perspectives of the ALC and LCA will be described below along with those of the ELCA and LCMS.

While the policies and procedures of these churches vary, it is generally the case that the materials under discussion here fall into one of two categories. Some (e.g., "Social Statements") represent the official policy of the church and are considered binding on staff members and others who may speak on behalf of the church. Others are merely presented to the church as resources for study and discussion. The reports of committees or commissions and texts prepared by church staff or outside authors are of the latter sort. Clearly, they are accorded less authority. It should not be surprising that churches in the tradition of Martin Luther and the Reformation principles of *sola fide* (faith alone) and *sola scriptura* (scripture alone) do not regard any of their social teachings as binding the consciences of individual members. In fact, the ELCA has decided that in its future social teaching statements it shall be recorded that the text was adopted by a majority of at least two thirds of the Churchwide Assembly. To any statements that elicit "significant division" in the assembly, an addendum shall be added summarizing dissenting points of view.

III. BIOETHICS IN THE LUTHERAN TRADITION

Despite their traditional suspicion of moral theology, the Lutheran churches in the United States have undertaken rather extensive exploration of issues in bioethics. As early as 1958 prominent Lutheran theologians were commissioned to consider Christian social responsibility in relation to contemporary problems. By the middle 1960s the LCA was publishing a series of booklets *Studies in Man, Medicine, and Theology*, in which theologians and physicians collaborated on such topics as "man as patient", "a physician's view of man", conception and contraception, abortion, and the dying patient [20]. When in the early 1970s bioethical controversy gained wide exposure in the media, Lutherans began to ask their churches for guidance. Aware that it hardly knew where to begin, the LCA held a consultation on bioethics in the spring of 1979. Among the seventy participants were physicians of various specialties, nurses, scientists, and lawyers, as well as theologians and church staff. Issues discussed in the conference papers included the role of theological reflection, the place of human values in scientific methodology, social regulation of scientific research, genetic engineering, human experimentation, and medical paternalism [39]. This consultation and a smaller follow-up conference in 1980 helped to determine the questions the LCA should consider and the manner in which they might be addressed most usefully. A similarly broad examination of bioethics is being planned by the LCMS. Doubtless, bioethics will continue to be a topic of study and conversation among Lutherans.

The remainder of this report is devoted to explication of the Lutheran churches' views on specific bioethical issues. Since the order of exposition follows the *Yearbook*'s convention rather than chronological order or the arrangement of topics in the documents themselves, some awkwardness and repetition is unavoidable. Accuracy in attributing the various texts necessitates an unfortunate alphabet soup of acronyms (ALC, LCA, LCMS, ELCA, LCUSA) and deference to ecclesiastical distinctions regarding degrees of authority (Social Statements, studies, convention resolutions, etc.). Readers are thanked for their patience.

A. *New Reproductive Technologies*

1. *Contraception*

In 1964 the LCA adopted the Social Statement "Marriage and Family" that called upon spouses "to exercise the power of procreation responsibly before God". Such responsibility implied "planning their parenthood". Expressing particular concern for "the health and welfare of the mother-wife", the Statement indicated that "irresponsible conception of children up to the limit of biological capacity and selfish limitation of the number of children are equally detrimental." It also advised that "choice as to means of conception control should be made upon professional medical advice" ([24], p. 243). "Marriage and Family" was actually the policy of one of the LCA's predecessor church bodies, and was embraced as an interim guideline while the LCA, then in its second year, developed its own position.

The first stage of this process was a study booklet by Carl Braaten, in which fellowship and children were affirmed as the two goods of marriage. Braaten judged unethical ("contrary to God's will and purpose for man") any acts which "violate the structural integrity of marriage as intended by God" ([2], p. 2). However, he saw "a vast difference between the couple who carefully plan when to have their children and the couple who selfishly refuse to have any at all" ([2], p. 9). Any effective form of contraception was justifiable in the first instance and none in the second. If, on the other hand, voluntary childlessness was sought for medical or genetic reasons and not for reasons amounting to a "selfish refusal", contraception was acceptable. While Braaten could not "say an unambiguous 'no' to sterilization once [he had recognized] the right of parents to decide never to have *more* children", he thought "it should not be used where other methods will accomplish the same result" ([2], pp. 25-26).

When the LCA finally elaborated its own official position in the 1970 Social Statement "Sex, Marriage, and Family", the ethical significance of contraception was again tied to the motivation of the couple within "a covenant of marital fidelity". Yet it was explicitly stated that "people have a right not to have children without being accused of selfishness or a betrayal of the divine plan; and every child has a right to be a wanted child". Moreover, the Statement asserted that "all persons are entitled to receive from governmental and voluntary agencies information about conception control". Responsible decisions about contraception might include "evaluation of such factors as the health of the potential mother, a reliable prognosis concerning the health of the possible child, the number and spacing of other children, the family's economic circumstances, and the rapid growth of the population" ([24], p. 279).

An ALC perspective was offered in a 1977 booklet *Health, Life and Death* (1977):

> Effective birth control methods facilitate responsible procreation and greatly enhance the ability to exercise stewardship of genetic resources. Enjoyment of sexual intercourse without fear of unwanted pregnancy is appropriate. Men and women are equally responsible for contraception and procreation ([60], p. 10].

Under certain circumstances it was conception that was "inappropriate" in the eyes of the ALC. These included conception "when without intent to carry the child to term, when a grossly defective infant is probable, when neither of the couple wants a child, when the parents are incapable of nurture, [or] when the result of rape or incest, or when induced by societal pressure" ([60], p. 10). Perhaps implicit in this observation was the judgment that when any of these were in prospect, contraception was obligatory. Voluntary sterilization was held to be "usually appropriate" ([60], p. 10).

The LCMS position, set forth in its "Human Sexuality: A Theological Perspective" (1981), concluded that:

> [i]n the absence of Scriptural prohibition, there need be no objection to contraception within a marital union which is, as a whole, fruitful. Moreover, once we grant the appropriateness of contraception, we will also recognize that sterilization may under some circumstances be an acceptable form of contraception...for couples who already have children and who now seek to devote themselves to the rearing of those children, for those to whom the birth of another child would be hazardous..., or for those who for reasons of age, physical disability or illness are not able to care for additional children. Indeed, there may be special circumstances which would persuade a Christian husband and wife that it would be more responsible and helpful to all concerned, under God, not to have children ([11], pp. 19-20).

However, the last observation was offered with the utmost caution.

2. *Artificial Insemination*

In 1967 Braaten declared "there is virtual consensus among Protestants that AIH (artificial insemination by the husband) is morally unobjectionable" ([2], p. 32). AID (artificial insemination by donor) was another matter. Nevertheless, after noting a host of prudential considerations which militate against it, Braaten concluded that the decision should be left to the couple. "We do not believe that Christian ethics can issue an unqualified 'Thou shalt not' on AID" ([2], p. 34). Ten years later an ALC document concurred. AID "may be perceived as appropriate for some married couples", though there are "moral and legal ambiguities that must be taken into consideration" ([60], p. 10). Further study and discussion was deemed necessary to determine the "propriety" of "sperm banks, surrogate gestation and genetic engineering".

The LCMS approved AIH – while noting that it "may be a way of avoiding underlying psychological problems within a marriage rather than treating them" – but rejected AID ([11], p. 38). In the latter, "the process of fertilization is removed from the personal context of the one-flesh union of husband and wife in a way that not even their consent can allow" ([11], p. 39).

The most recent and extensive Lutheran discussions of AID (and the other reproductive issues treated below) appeared in a series of seven essays published by the LCA in 1986 under the title "Procreation Ethics" [41]. Each was the work of a single author who was free to analyze the issues as he or she saw fit. Longer versions of the essays were published in *Questions About the Beginning of Life*, edited by Edward D. Schneider and published by the ALC's Augsburg Publishing House in 1985 [48]. Both publications were products of a joint LCA-ALC study project. In the essay on artificial insemination, Edward D. Schneider argued that AID is "not an ethically acceptable alternative", because it "separates procreation from marriage and thereby violates the marriage covenant" [49], p. 12). While AIH was unobjectionable, AID was said to have grave psychological risks for the couple and required the donor to "exercise his procreative powers apart from any marital bond or commitment". His function was that of "a sperm salesman, failing to take full responsibility for his biological offspring". While AID did not "fall into the category of some unforgivable sin", Schneider advised the infertile "to come to terms with their childlessness or to seek children through adoption" ([49], p. 13).

3. *In Vitro Fertilization*

In the early 1980s the Lutheran Council in the USA (LCUSA), a cooperative agency of the LCA, ALC, and LCMS, undertook a study of in vitro fertilization (IVF) [13]. Their document set forth the "strongest case" arguments for and against IVF. The unanimous finding of the 11-member committee was that "IVF does not *in and of itself* violate the will of God as reflected in the Bible, when the wife's egg and husband's sperm are used" ([13], p. 31). No consensus was reached on the use of donor sperm or eggs or on whether all fertilized eggs must be implanted. Despite their expressed caution concerning psychological problems, early resort to IVF before exploring alternatives, the cost of the IVF process, and its impact on the marriage, the LCUSA committee concluded "that marriage partners may in good conscience consider using IVF to conceive and give birth to a child" ([13], p. 31).

Paul T. Jersild, a member of the LCUSA study group, wrote the IVF essay for the LCA-ALC series [22]. Finding no reason for the church to reject IVF categorically, he advocated public policies restricting IVF "to married couples who desire a child of their own lineage", limiting embryo storage to facilitating second and third transfers for such couples. Concerned that society might not have the moral consensus necessary to support such restrictions, Jersild found it imperative "to resist those inclinations that turn human life in its early stages into a product designed to satisfy a consumer-oriented society" ([48], p. 48).

4. *Surrogate Motherhood*

Janet Dickey McDowell, writing in the LCA-ALC series, held that "Christian perceptions of the significance of human procreation and its place within the marital relationship are not compatible with the basic premise of surrogate motherhood: that one could deliberately conceive and bear a child with no commitment either to the child or its father" ([34], p. 10). It is sometimes argued in defense of surrogacy that surrogates provide "the gift of life", but McDowell observed that "a Christian understanding of procreation does not view children as entities to be created in order to be bestowed on others, as though they were handmade sweaters or cookies". Nevertheless, she concluded that "wholesale condemnation may not be appropriate", and that it should be "left to individual conscience" to judge whether surrogacy's benefits "outweigh the distortion of procreation, especially for the surrogate, that makes those benefits possible".

5. *Genetic Manipulation*

On June 8, 1983, the Reverend James R. Crumley, Jr., Presiding Bishop of the LCA, joined 57 other religious leaders in a resolution calling upon Congress to prohibit genetic engineering of the human germline cells. The group included 21 Roman Catholic bishops and the heads of several other Protestant denominations. The initiative was organized by Jeremy Rifkin of the Foundation on Economic Trends, and was accompanied by a "Theological Letter Concerning the Moral Arguments". Crumley, who had headed a panel that studied genetic engineering for the National Council of Churches, told the press that he fully supported only the resolution. "There are some aspects of genetic therapy [for human diseases] that I would not want to rule out", he said. "My concern is that someone would decide what is the most correct human being and begin to engineer the germline with that goal in mind" [7].

Sharing Crumley's concern but rejecting such calls for prohibition was James H. Burtness, author of the LCA-ALC pamphlet *Genetic Manipulation*. Burtness wrote, "New tools for dealing with nature may be the means by which God is using us to work at bringing back the whole created order into line with God's own purpose. The Christian is *inclined*, therefore, to underscore the promise rather than the peril of new discoveries" ([4], p. 10). The potential benefits and the fact that it has proved possible, so far, to regulate procedures and to control possible hazards led Burtness to place the burden of proof on the opponents of genetic manipulation. Participation in the management of creation is something which Christians accept "as both gift and responsibility", yet, because of their "awareness of the demonic potential of human activity, [Christians] will insist that regulation is necessary" ([4], p. 11).

6. *Genetic Screening and Counseling*

Endorsing the use of amniocentesis in 1977, the ALC cautioned that it must be "questioned as a routine screening procedure, as a means of assuring the desired sex of the offspring, when used against the wishes of a parent, or when abortion is the only option offered" ([60], p. 11). Its use raised two troubling questions: (1) "Is there such a thing as a life not worth creating? (2) Is there such a thing as a life not worth living?". The ALC document saw that "decisions about abortion and the minimum acceptable quality of human life must evolve from answers, or partial answers, to these questions". But no attempt was made to offer these answers. The ALC dismissed popular concerns about the degradation of the human gene pool, and deplored the fact that "in the past, legal and societal sanctions in the name of eugenics have usually degenerated into racial, ethnic, economic and social criteria of human worth". The ALC did, however, acknowledge the church's obligation "to foster genetic education" and "to assist [those at risk] in obtaining adequate expert genetic counseling". These efforts were seen as an endorsement of "responsible parenthood".

The LCA-ALC booklet, by James M. Childs, Jr., emphasized that genetic screening should not be viewed as "interference with God's plan" ([6], p. 6). Rather, "responsible use of our genetic knowledge in screening and counseling for the prevention of unnecessary suffering is in accord with God's purposes". At the same time, Childs warned, "it is also possible that sinful human beings will misuse the knowledge they gain or be misguided in their application of that knowledge" ([6], p. 7). Stigmatization of minority populations, coercive legal restrictions on marriage, invasion of the right to privacy, and a lessening of acceptance for those afflicted with birth defects are very serious, even if sometimes exaggerated, concerns which require safeguards. Though Childs felt "it necessary to oppose mandatory public screening in order to protect freedom of choice in procreation", such freedom should not be exercised without restraint. "Christian couples facing the risk of severe genetic disorders in their children need to consider their obligation to forego the freedom to procreate, and not gamble with the odds in order to prevent unnecessary suffering" ([6], pp. 8-9).

7. *Prenatal Diagnosis*

In the LCA-ALC series, Edmund N. Santurri observed that the advent of prenatal diagnostic techniques raised serious moral questions. For example, given that amniocentesis involves some degree of risk for the fetus, "is taking such risks warranted by the concern to prevent *possible* genetic defect or to prepare a family for the *possible* birth of a child with genetic disease?" ([47], p. 6). Aware that there are no easy or clear answers and that different probabilities would lead to different judgments, Santurri did not presume to settle the questions. Judging that there were cases (e.g., Tay-Sachs and Lesch-Nyhan syndrome) where the strong Christian "presumption against fetal destruction" might be set aside and genetic abortion could be "regarded intelligibly as [an] expression of parental care", Santurri insisted that in all cases "morally responsible judgments will be those which keep in full view the fetus' well-being" ([47], pp. 10-11). In some tragic circumstances, abortion might be in the interests of the child. However, "it would be implausible to suggest that for a typical Down's syndrome child life is not worth living, and thus difficult to construe an abortion of a Down's fetus as an informed expression of parental care".

Santurri also explored the moral issues raised by mass screening programs to detect neural tube defects in the general population. He argued that a just public policy would assure that all parents at risk would have equal opportunity for entry. And yet he felt that "there ought to be reservations about such goals because there ought to be reservations about the abortions which the program inevitably encourages". Such moral ambivalence, Santurri concluded, "will be experienced by those who believe both that fairness requires equal access to the services society affords, and that caring for children, prenatally or postnatally, is an activity which signifies in the created order the divine providential economy at the heart of the Christian story" ([47], pp. 12-13).

B. *Abortion*

In the Lutheran community, as in society at large, no issue is more painful, provocative, and divisive than abortion. In response to an informal readers' poll in the summer of 1989, the ELCA denominational magazine *The Lutheran* received 320 responses in favor of laws restricting access to abortion and 296 in opposition [37]. The ELCA has a task force at work preparing a draft of a "Social Teaching Statement" for consideration at a Churchwide Assembly in the summer of 1991. To appreciate fully the challenge before the ELCA, it is necessary to recall the positions of its predecessor churches, the ALC and the LCA, as well as that of the LCMS.

1. *The Lutheran Church-Missouri Synod*

The LCMS adopted a resolution in 1971 declaring "willful abortion as contrary to the will of God" ([30] Res. 2-39). A report of the Commission on Theology and Church Relations made it clear that by "willful abortion", the LCMS had in mind those abortions which are non-therapeutic, i.e., those performed for some reason other than "to safeguard the health or life of the patient, or to prevent the birth of a severely crippled, deformed or abnormal infant" ([9], p. 1). The Commission observed that "not wanting to be a mother does not provide a proper justification", and that "rape or incest create very special problems requiring pastoral counseling of the most sensitive kind". No judgment was made concerning the latter cases. Since 1971 several conventions of the LCMS have reiterated this position. Resolutions have called upon members "to protest publically the sin of abortion", and to support various legislative efforts "to provide protection for all human life", including amendment of the United States Constitution ([30], 1977 Res. 308c, 1979 Res. 3-02A, 1981 Res 3-02). In addition to participating in a number of LCMS programs designed to implement the church's teaching, members were urged to become involved in "alternatives to abortion" programs and in Lutherans for Life, an independent anti-abortion organization whose members are drawn from all the Lutheran churches ([30], 1977 Res. 3-08c, 1979 Res. 3-02A).

The theological rationale for the LCMS position was set forth in a 1984 study booklet *Abortion in Perspective*. Granting that "the Scriptures do not specify the moment at which a new individual human being comes into existence", the LCMS maintained that science and medicine point to fertilization ([12], p. 27). Nevertheless, four "scriptural principles" were said to shape Christian belief, character, and action so as to preclude acceptance of abortion:

1. Human life, at every stage of its development, is valued by God.
2. Human lives are entrusted by God to our care.
3. There are limits to human freedom.
4. Moved by their hope in God, Christians must be a people glad to receive children into the human family ([12], pp. 27-32).

On the basis of these principles, the LCMS regarded "abortion on demand as not only a sin against the Fifth Commandment forbidding the destruction of human life, but also as a grievous offense against the First....The act of abortion clearly manifests a refusal to honor God as the Creator and to seek Him above all else in time of need" ([12], p. 32).

In its "ethical reflections" section, the LCMS booklet outlined the church's anti-abortion argument:

Major Premise: The lives of human beings – whatever their stage of development or achievement – are entitled to equal care and protection.
Minor Premise: The unborn child is a human being.
Conclusion: The life of the unborn child is entitled to equal care and protection ([12], p. 36).

The LCMS believed that the expansion of scientific knowledge of fetal development and progress in fetal therapy lend increasingly persuasive force to the minor premise. Our society's rejection of the conclusion suggested to the LCMS that support for the major premise had begun to erode and that the erosion was likely to continue.

The LCMS was aware, of course, that the articulation of moral judgments is not a substitute for "the delicate administration of Law and Gospel to those troubled by decisions related to abortion" in a pastoral context ([12], p. 37). LCMS documents offer guidance for pastors, counselors and others who may offer "spiritual care".

Reaffirming earlier LCMS calls for the participation of Lutheran citizens in political efforts to restrict abortion, *Abortion in Perspective* spoke cautiously about initiatives for a constitutional amendment. It placed its "greatest hope" in the possibility that the Supreme Court would "extend greater protection to unborn human beings" ([12], p. 43). In 1989 an LCMS convention welcomed the *Webster* decision "as a necessary first step toward the full restoration of the right to life" ([30], Res. 3-09A).

2. *The Lutheran Church in America*

In marked contrast to the position of the LCMS, the LCA was reticent to address abortion issues, and when it did speak, it was generally "pro-choice". The LCA's official policy on abortion was established in the 1970 Social Statement "Sex, Marriage and Family". At that time abortion was illegal in many states, and the LCA was asked to provide guidance to pastors ministering to people who were considering either an illegal abortion or travel to a state in which abortion was legal. From this pastoral perspective, the statement declared that, "On the basis of the evangelical ethic, a woman or couple may decide responsibly to seek an abortion" ([52]; [24], p. 279). In other words, the possibility that, in some circumstances, abortion might be morally acceptable could not be precluded. While the Statement suggested that "earnest consideration should be given to the life and total health of the mother, her responsibilities to others in her family, the stage of development of the fetus, the economic and psychological stability of the home, the laws of the land, and the consequences for society as a whole", it did not offer an exposition of the "evangelical ethic" or demonstrate the reasoning by which certain abortions might be justified. The Statement reported the conclusion of an argument without supplying the argument itself. This was unfortunate, because in this context the argument would have been more important than the conclusion. In any event, the mutual counsel of the Christian community was understood to be the context of the evangelical ethic.

Over the years, as abortion became more readily available, many persons within the church were distressed by what they perceived to be a growing moral complacency and a diminution of the sense in which even justifiable abortions are tragic. Also, concerns were expressed about the extent to which a Christian understanding of human nature was being displaced by an atomistic individualism. Individuals and groups within the LCA were active on all sides of various political and legislative controversies. Some were moved to ask where the LCA's 1970 Statement stood with respect to current political slogans, cultural notions, and societal trends. In response to these questions, the 1978 convention of the LCA adopted a "Clarifying Minute on Abortion", which declared that the 1970 Social Statement "opposes" the concept of "abortion on demand" and the "use of abortion as an alternative form of contraception" ([29]; [24] p. 281). While these assertions expressed the wishes of a majority of the delegates in 1978, whether they "clarified" or were even consistent with the 1970 Social Statement was open to serious question.

The LCA did produce some study material on abortion. Franklin Sherman's *The Problem of Abortion After the Supreme Court Decision* (1974) was intended to assist readers in thinking about abortion as a moral rather than simply a legal problem. "The Christian ethic", Sherman explained, "will add to the language about rights a discussion of...reverence for life and responsible freedom" ([50], p. 13). But the discussion in the booklet yielded no "clear and simple answer". Abortion was "a complex problem, requiring a delicate balancing of factors" ([50], p. 23). Ten years later the LCA opted for a different pedagogical approach when it published a collection of four essays reflecting – collectively, though not individually – the diversity of theological and ethical judgments within the LCA [40]. The title *Abortion and Christian Conscience* was chosen to refocus the church's attention on the question, "How should I, as a Christian, think about abortion for my life?". In the same year, 1984, the LCA decided not to bring to the floor of its convention a new abortion resolution (drafted by staff), which would have reopened consideration of the LCA's policy. The issue was perceived to be too hot to handle, and the position taken in 1970 (and "clarified" in 1978) remained the LCA's official policy ([24], p. 114).

3. *The American Lutheran Church*

The position of the ALC occupied a middle ground between the policies of the LCMS and the LCA. In "Christian Counseling on Abortion", a 1974 Statement of Comment and Counsel, the ALC declared itself to be "pro-life", rejecting "induced abortion as a ready solution to problem pregnancies" [16]. Nevertheless, it recognized that "an induced abortion may be a necessary option in individual human situations". A characteristically Lutheran theme was sounded when the document announced that, inasmuch as the church was "deeply aware of the sinfulness in every human decision" and "deeply confident of God's grace and forgiveness", it had "no need to itemize a list of circumstances under which abortion is acceptable or is forbidden".

In 1976 the ALC received as information two documents: a general statement "The Value of Human Life" [17], which expressed in passing "a clear moral preference...against abortion", and *A Statement on Abortion* written by James H. Burtness [3]. The statement was a personal one and had the same status as the LCA study booklet by Franklin Sherman. Burtness could not escape the conclusion that "the current and widespread practice of totally unregulated abortion is morally intolerable....The most important question...is...how to effectively regulate abortion so that it is perceived to be the very serious (and only occasionally necessary) moral tragedy that it is" ([3], p. 15).

The ALC's official policy was established in a 1980 "Statement of Judgment and Conviction". Here the church affirmed that "human life from conception...is always sacred" and "that induced abortion ends a unique human life" [19]. Acknowledging that "there may be circumstances when...an induced abortion may be a tragic option", the ALC rejected "the practice in which abortion is used for personally convenient or selfish reasons". It deplored "the alarming increase of induced abortions" since 1973, and "the absence of any legal protection for human life from the time of conception to birth".

4. *The Evangelical Lutheran Church in America*

When the ALC and LCA joined together in 1988 they bequeathed to the new ELCA the considerable challenge of reconciling their disparate stands on abortion. At the ELCA's first Churchwide Assembly in August 1989, the long simmering issue boiled over. The convention adopted an omnibus resolution on "reproductive issues" which called on pastors "to encourage free access to services". When delegates were informed that "services" included abortion, they voted to reconsider, and eventually removed the controversial language. The headline in *The New York Times*, "Lutherans...Take a Neutral Stand on Abortion", was inaccurate, in that the convention's several days of heated wrangling demonstrated that many of the delegates were by no means "neutral" [31]. When the ELCA reassembles in 1991, the abortion issue will be on the front burner. At that time the convention is scheduled to consider a proposed "Social Teaching Statement" on abortion now being drafted by a 15-member task force. However, as of June 1990, the panel was reported to be deadlocked on abortion for socio-economic reasons, for the mother's "emotional health", and where there is "serious deformity" in the fetus. Nor could they agree on what kind or degree of governmental regulation is appropriate, or even on a norm of sexual abstinence outside of marriage. To date they have agreed only that post-viability abortions are to be avoided, and that in cases of rape, incest, or mortal threat to the mother, abortion may be a "tragic option" [14]. All of this suggests that the Lutheran community remains deeply divided over the issue of abortion.

C. *Care of Severely Handicapped Newborns*

Hans O. Tiefel argued in the LCA-ALC *Procreation Ethics* series that "our reasons for stopping treatment of handicapped newborns must hold across the lifespan and must be compatible with our reasons *for* treating" ([67], p. 11). This was in keeping with the LCA's 1982 Social Statement "Death and Dying", which was intended to apply to newborns as well as adults ([57]; [24], pp. 203-211). Tiefel's particular concern was to challenge the individualistic ethos of American society, which is reluctant to ascribe rights to such weak, utterly dependent beings, and to underwrite the "governmental support essential to create and finance [their] continuing care" ([67], p. 10). The "rights" of handicapped newborns were not the primary issue from a Christian perspective. "We owe them more than their rights, whatever those be....The Church has always known that we are called to save, to heal and literally to redeem human lives whenever possible". And yet, Tiefel recognized,

> the very preciousness of these children may compel us to cease treating when our intervening will no longer do them any good, when our therapies become fruitless and therefore harmful. We must cease when life and continued treatment become so burdensome for those infants that our best efforts will not help to make them better. We must stop when the continued treatment is fruitless and becomes overtreatment, and thereby mistreatment ([67], p. 11).

In certain cases (e.g., anencephaly or Lesch-Nyhan syndrome), treatment may be "senseless from the start". This reasoning, Tiefel observed, "does not single out unwanted infants as nonpersons", regarding their lives as "dispensable or replaceable", nor does it "prefer the rights or values of adults to the lives and needs of these patients".

D. *Consent to Treatment*

"To treat [a competent patient or an incompetent who made his or her preferences known while still competent] in violation of his or her deeply held, carefully considered, and clearly expressed preferences is to do violence to that person", declared the LCA's 1982 Social Statement "Death and Dying" ([57]; [24], p. 207). While the LCA regarded stewardship of life, one's own or another's, as a fundamental moral value, it allowed competent patients "in good conscience to refuse burdensome treatments in some situations". Even where a patient's refusal may seem to be a dereliction of responsible stewardship, the LCA was prepared to respect his or her autonomy. However, in cases where competency is diminished due to pain or to other factors, or where expressions of preference may not represent a person's true wishes, the LCA suggested that "it may be appropriate to administer treatment (by authority of court order, if necessary) if doing so would sustain the life of the patient".

E. *Equitable Access to Health Care*

In the mid 1970s, and again in 1983, the LCA sponsored a conference on health care in America. The first resulted in the publication of *Health Care in America: A National Illness* [27]. A contributor to the volume concluded that the "paradox of affluence and inadequacy of health care", particularly among the poor and non-white, posed a central question:

> Are we ready as a nation to accept as a social and political purpose that quality health care services should be accessible to all persons in their own communities without regard to their ability to pay? This is a question that each of us must help to answer and must do so in response to our Christian concept of personhood and the relationship between persons. Within these constraints an affirmative answer seems inevitable ([27], p. 51).

As a binational church, the LCA was able to draw upon the experience of its Canadian members in studying a variety of models for governmental involvement in health care delivery.

The American Lutheran Church was also interested in a more just distribution of health care. Study material prepared by the ALC's Task Force on Ethical Issues in Human Medicine in 1977 encouraged "the establishment of basic standards of curative and preventive health care and the adoption of effective control measures for cost and quality of such care" ([60], pp. 6-7). Here the concern for equitable access to medical services was linked to "a sense of Christian stewardship" in "our demands upon and usage of these services". Government, private agencies, and consumers were called upon to "exercise imagination and creativity in developing innovative health care systems".

While the studies quoted above were not the churches' official policy, nor did they discuss specific legislative options, they did establish the church's interest in expanding access to health care. This interest found expression in a 1978 LCA Social Statement "Human Rights: Doing Justice in God's World", which asserted that "older adults should not be denied, because of circumstances beyond their control, adequate housing, sustenance, or health care" ([55]; [24], p. 233). What may have been implicit in 1978 became explicit in the 1980 LCA Social Statement "Economic Justice: Stewardship of Creation in Human Community". Here access to health care was identified as "substantive entitlement" for which government was responsible ([55]; [24], p. 218).

F. *Death and Dying*

1. *Theological Perspectives*

Within Lutheranism, as in the Christian tradition as a whole, death has been imaged in a variety of contrasting and complementary ways. The LCA's 1982 Social Statement "Death and Dying" identified five. Death was viewed "as a natural part of the life cycle", "as an unwelcome event that has a tragic dimension", "as a deliverance" from prolonged suffering "and in this sense as a friend", as "an enemy" whose "sting" is attributable to "the alienation and estrangement" from God "pervasive in our lives", and as that which is overcome in "the New Testament message of Christ's victory over death" in the resurrection. However death is viewed, the Statement declared, "all who experience death-and dying situations can be certain of God's love" ([57]; [24], p. 203-204).

The reference to death "as a friend" was highly controversial, and in two years an LCA convention formally directed that "friend" be changed to "merciful" ([24], p. 211). This was the only time in the 25-year life of the LCA that one of its 19 Social Statements was later amended. But not even this substitution would have satisfied the authors of a 1979 LCMS study guide, which asserted that

> God created human beings to live and not to die. Death in any form is inimical to what God originally had in mind for his creation....To speak of 'death with dignity' or 'merciful release,' therefore, consists of engaging in unholy rhetoric....Had it not been for man's own rebellion...there would be no death ([10], p. 18).

Despite theological disagreement over the interpretation of Scripture, the Lutheran churches are in general agreement concerning the ethics of death and dying.

The LCA Social Statement identified some "interpretive principles" which should inform Christian decision-making:

a) Life is a gift of God, to be received with thanksgiving.
b) The integrity of the life processes which God has created should be respected; both birth and death are part of these life processes.
c) Both living and dying should occur within a caring community.
d) A Christian perspective mandates respect for each person; such respect includes giving due recognition to each person's carefully considered preferences regarding treatment decisions.
e) Truthfulness and faithfulness in our relations with others are essential to the texture of human life.
f) Hope and meaning in life are possible even in times of suffering and adversity - a truth powerfully proclaimed in the resurrection faith of the church ([24], pp. 204-205).

The competent patient was considered "the prime decision-maker". Whenever patients were incapable of deciding for themselves and no expression of preferences was on record, family, physicians, pastor, and close friends could seek consensus. The LCA recognized a role for hospital ethics committees, but discouraged "appeal to the courts...unless so doing is the only way to protect individual rights or to resolve the controversy" ([24], p. 205).

2. *Withdrawing or Withholding Treatment*

The LCA document indicated how the above principles might be applied in specific cases. Its judgments were consonant with those of other Lutheran groups. Three distinct types of situations were addressed.

a. *The Irreversibly Dying Person*

When disease progresses to the point where no effective therapy is available and the patient is in "the final stages of the dying process", it is appropriate to refrain from resuscitation and discontinue the use of life support systems. "To try desperately to maintain the vital signs of an irreversibly dying person for whom death is imminent is inconsistent with a Christian ethic that mandates respect for dying, as well as for living" ([24], p. 205). The ALC agreed, calling it a "grave injustice" ([60], p. 18). The LCMS found that where "there is no real hope for recovery even with life support" in the professional judgment of physicians, "a Christian may in good conscience 'let nature take its course'" ([10], p. 28). However, "supportive care intended to maintain comfort and otherwise respond to the needs of the dying person" remains appropriate.

b. *Burdensome Treatments*

Where there are forms of therapy which may sustain life but which themselves involve considerable discomfort, a choice must be made between quantity of life and quality of life for the patient. Recognizing the particularity and extreme difficulty of such situations, the LCA did not attempt to adjudicate them. Instead it offered "factors to be considered" and recommended second opinions and consultations. The factors were:

1. the probability that...treatment will sustain the life of the patient;
2. the length of time that the life of the patient is likely to be sustained;
3. the anticipated risks and side effects;
4. other forms of treatment available, if any, and their relative advantages and disadvantages;
5. the patient's adjustment to hospitalization or to the treatment;
6. the extent to which the treatment will interfere with the person's most cherished activities; and
7. available support systems at home or in alternative institutional settings ([24], p. 206).

Obviously, the LCA assumed that some treatments, though potentially effective, may be so burdensome that they may be withheld or withdrawn.

c. *Chronically Ill Individuals*

The LCA emphasized that "the foregoing should not be taken to imply that chronically ill persons should be allowed to die because their lives are judged to be not worth living or because they are viewed as burdensome or useless to society. Whether the patient is a newborn infant...or an aged person..., the Christian response in such cases must be a strong presumption in favor of treatment". The presumption was rebuttable "in cases of extreme and overwhelming suffering from which death would be a merciful release, or in cases in which the patient has irretrievably lost consciousness" ([24], p. 206).

3. *Use of Pain-Killing Drugs*

Both the LCA and the LCMS concurred; in the words of the LCMS document, "administering pain-killing medications, even at the risk of shortening life, is permissible, since this does not entail the choice of death as either a means or an end" ([103], p. 29). The LCA added that "adjustments in administering such drugs should be made so as not to deprive the patient of consciousness prematurely" if possible, but in any case "recognition should be given to patient preferences, when they are known" ([24], pp. 207-208).

4. *Active Euthanasia*

The LCA and LCMS joined the ALC in recognizing "a distinct moral difference between killing and allowing to die" ([60], p. 20). In 1979 the LCMS declared that "[e]uthanasia, in its proper sense, is a synonym for mercy killing, which involves suicide and/or murder. It is, therefore, contrary to God's law" ([10], p. 28). According to the LCA, "Christian stewardship of Life...mandates treasuring and preserving the life which God has given", one's own or another's "meaning and hope are possible in all life's situations, even those involving great suffering. To depart from this view by performing active euthanasia, thereby deliberately destroying life created in the image of God, is contrary to Christian conscience" ([24], p. 208).

5. *Definition of Death*

In 1977 an ALC document observed that "definitions of death consist of more than biological facts. They must also consider the personal and spiritual dimensions of life" ([60], p. 18). While it is not clear what the authors intended by this, it is unlikely that they meant to endorse an "upper/higher brain" definition. Two years later an LCMS, study observed without comment that "a new criterion (brain death) for death is being used in medicine today" ([10], p. 15). It pointed out, however, that patients "comatose and morbid on admission to the hospital" sometimes recover fully. Therefore, "a patient who has suffered brain damage but who possesses a spontaneously beating heart, active circulation, active respirations and who produces urinary output must be sustained and observed carefully while monitoring determines whether other systems are involved or become involved with the consequence that total death ensues" ([10], p. 14). The LCA 1982 Social Statement did not address this issue, but it received extended treatment in the accompanying study guide. Of the alternatives then under discussion, the "whole/total brain" definition was held to be the most adequate. The author, Daniel E. Lee, observed that it is best to be forthright about a decision to allow a still living person to die, "rather than obscuring it in vague language that blurs the distinction between discontinuing a life support system because someone has died, or discontinuing it so that someone may be allowed to die" ([26], p. 53).

According to the LCMS "often the time prior to death is so wrapped in mystery that no one ought forcibly to interrupt the movement of man's spirit as it may be communicating through God's Spirit with his Creator and Redeemer by way of responding in trust and inner yearning" ([10], p. 29). However, it was not indicated what sort of acts would constitute "forcible interruption", or what might be the implications in regard to the definition or determination of death.

G. *Organ Donation*

The 1982 LCA Social Statement encouraged consideration of "the possibility of organ donation as a means of sharing life with others, authorizing an autopsy the donation of the body for scientific purposes" ([24], p. 210). Two years later an LCA convention adopted a resolution that went even further. It recognized that "the donation of renewable tissue and live organs can be an expression of sacrificial love for a neighbor in need". Members were encouraged to communicate their wishes to their families, physicians, pastors, and health care institutions to make the necessary familial and legal arrangements in advance. The use of signed donor cards was recommended. Pastors and church agencies were urged to facilitate donation. The resolution also called upon government to encourage voluntary donation, discourage coercive donation, disallow the sale and purchase of organs, and to "assure the efficient, equitable distribution of human organs and tissues for transplants" [29].

As early as 1965 the LCMS found it unnecessary to study the practice of bequeathing human bodies for medical research, because it regarded it as "a matter of individual conscience" [30]. In 1981 the LCMS adopted a resolution calling upon the church to implement programs to promote donation, and encouraged "family members to become living kidney donors" [30].

H. *Conclusion*

The studies surveyed above represented a substantial and serious engagement with the problems and challenges of modern medicine. Whatever their intrinsic significance as statements of church policy, their greatest value lay in the reflection and debate occasioned by their drafting and adoption, as well as by their favorable or unfavorable reception in seminaries, health care institutions, and parishes throughout the churches. In this way they have helped to make the Lutheran churches "communities of moral discourse".

It is ironic that a theological tradition with such deep reservations about moral casuistry should have undertaken such extensive examinations of bioethical issues. Perhaps it is simply the case that today such efforts are recognized as constituent to "faith working through love" in service to the neighbor. When neighbors are confronted with infertility, abortion, inadequate health care, sickness, and suffering as death draws near, or perplexed by the ways in which our astounding medical and technological capabilities so often yield a mix of weal and woe, Christian moral reason (even in its Lutheran varieties) is obliged to take up the practice of bioethics.

Wittenberg University
Springfield, Ohio, U.S.A.

BIBLIOGRAPHY

1. Bonhoeffer, D.: 1955, *Ethics*, ed. E. Bethge, trans. N.H. Smith, Macmillan, New York.
2. Braaten, C.: 1967, *The Ethics of Conception and Contraception*, The Lutheran Church in America, New York.
3. Burtness, J.: 1976, *A Statement on Abortion*, The American Lutheran Church, Minneapolis.
4. Burtness, J.: 1986, *Genetic Manipulation*, The Lutheran Church in America, New York.
5. Burtness, J.: 1989, "When Christians Disagree", *The Lutheran* (August 9), 12-14.
6. Childs, J.: 1986, *Genetic Screening and Counseling*, The Lutheran Church in America, New York.
7. "Clerics Ask Ban on Gene Tampering": 1983, *Newsday* June 8, p. 6.
8. Clouser, K.: 1973, "The Sanctity of Life: An Analysis of a Concept", *Annals of Internal Medicine* 78, 119-125.
9. Commission on Theology and Church Relations: 1971, *Abortion: Theological, Legal and Medical Aspects*, The Lutheran Church-Missouri Synod, St. Louis.
10. Commission on Theology and Church Relations: 1979, *Report on Euthanasia with Guiding Principles*, The Lutheran Church-Missouri Synod, St. Louis.
11. Commission on Theology and Church Relations: 1981, *Human Sexuality:*

A Theological Perspective, The Lutheran Church-Missouri Synod, St. Louis.
12. Commission on Theology and Church Relations: 1984, *Abortion in Perspective*, The Lutheran Church-Missouri Synod, St. Louis.
13. Division of Theological Studies: 1983, *In Vitro Fertilization*, Lutheran Council in the USA, New York.
14. "ELCA Abortion Task Force Cites Viability as Gauge": 1990, *The Lutheran* 3:11 (August 8), 32.
15. Enquist, R. H. Wildberger: 1967, *Christian Faith and the Dying Patient*, Lutheran Church in America, New York.
16. General Convention: 1974, *Christian Counseling on Abortion*, The American Lutheran Church, Minneapolis.
17. General Convention: 1976, *The Value of Human Life*, The American Lutheran Church, Minneapolis.
18. General Convention: 1980, *Human Sexuality and Sexual Behavior*, The American Lutheran Church, Minneapolis.
19. General Convention: 1980, *A Statement of Judgment and Conviction*, The American Lutheran Church, Minneapolis.
20. Heinecken, M. (ed.): 1967, *Man, Medicine and Theology*, Lutheran Church in America, New York.
21. Jenson, R.: 1967, "Man as Patient", in M. Heinecken (ed.), *Man, Medicine and Theology*, Vol. 1, Lutheran Church in America, New York.
22. Jersild, P.: 1986, *In Vitro Fertilization*, The Lutheran Church in America, New York.
23. Kersten, L.: 1970, *The Lutheran Ethic: The Impact of Religion on Laymen and Clergy*, Wayne State University Press, Detroit.
24. Klein, C. C. von Dehsen: 1989, *Politics and Policy: The Genesis and Theology of Social Statements in the Lutheran Church in America*, Fortress Press, Minneapolis.
25. Lazareth, W.: 1986, "Lutheran Ethics", in J. Childress (ed.), *The Westminster Dictionary of Christian Ethics*, Westminster Press, Philadelphia.
26. Lee, D.: 1983, *Death and Dying: Ethical Choices in A Caring Community*, Lutheran Church in America, New York.
27. Letts, H. (ed.): 1974, *Health Care in America: A National Illness*, Lutheran Church in America, New York.
28. Løgstrup, K.: 1971, *The Ethical Demand*, trans. T. Jensen, Fortress Press, Philadelphia.
29. Lutheran Church in America: 1984, *Minutes*, New York.
30. Lutheran Church-Missouri Synod: 1965-1989, *Convention Resolutions*, St. Louis.
31. "Lutherans, Reversing Earlier Vote, Take a Neutral Stand on Abortion": 1989, *New York Times*, August 31.
32. MacIntrye, A.: 1984, *After Virtue*, 2nd ed., University of Notre Dame

Press, Notre Dame, Indiana.
33. Marty, M.: 1983, *Health and Medicine in the Lutheran Tradition*, Crossroad, New York.
34. McDowell, J.: 1986, *Surrogate Motherhood*, The Lutheran Church in America, New York.
35. Meilaender, G.: 1987, *The Limits of Love: Some Theological Explorations*, The Pennsylvania State University Press, University Park, Pennsylvania.
36. Meilaender, G.: 1989, "Abortion: The Right to an Argument", *Hastings Center Report* 19:6, 13-16.
37. Miller, D.: 1989, "The Great Divide", *The Lutheran*, 2 (August 9), 10-11.
38. Miller, D. and Delloff, L.: 1990: "Special Section: A Right to Die?", *The Lutheran* 3 (August 8), 6-18.
39. Nelson, P. (ed.): 1982, *Collected Papers in Bioethics*, Lutheran Church in America, New York.
40. Nelson, P. (ed.): 1984, *Abortion and Christian Conscience: Theological and Ethical Perspectives*, Lutheran Church in America, New York.
41. Nelson, P.: 1986, *Procreation Ethics*, The Lutheran Church in America, New York.
42. Neuhaus, R.: 1990, "The Way They Were, The Way We Are: Bioethics and the Holocaust", *First Things* 1, 31-37.
43. Niebanck, R.: 1980, *Economic Justice: An Evangelical Perspective*, The Lutheran Church in America, New York.
44. Office of Church in Society: 1977, *Death and Dying*, The American Lutheran Church, Minneapolis.
45. Outka, G.: 1974, "Social Justice and Equal Access to Health Care", *The Journal of Religious Ethics* 2, 11-32.
46. Rendtorff, T.: 1989, *Ethics*, Vol. 2, *Applications of an Ethical Theology* trans. K. Crim, Fortress Press, Minneapolis.
47. Santurri, E.: 1986, *The Ethics of Prenatal Diagnosis*, The Lutheran Church in America, New York.
48. Schneider, E. (ed.): 1985, *Questions About the Beginning of Life*, Augsburg Publishing House, Minneapolis.
49. Schneider, E.: 1986, *Artificial Insemination*, The Lutheran Church in America, New York.
50. Sherman, F.: 1974, *The Problem of Abortion After the Supreme Court Decision*, Lutheran Church in America, New York.
51. Siirala, A.: 1964, *The Voice of Illness*, Fortress Press, Philadelphia.
52. Social Statement: 1970, "Sex, Marriage and Family", The Lutheran Church in America, New York.
53. Social Statement: 1972, "The Human Crisis in Ecology", The Lutheran Church in America, New York.
54. Social Statement: 1978, "Aging and the Older Adult", The Lutheran Church in America, New York.
55. Social Statement: 1978, "Human Rights", The Lutheran Church in

America, New York.
56. Social Statement: 1980, "Economic Justice", The Lutheran Church in America, New York.
57. Social Statement: 1982, "Death and Dying", The Lutheran Church in America, New York.
58. Stortz, M.: 1989, "Where Can We Stand?", *The Lutheran* 2 (August 9), 6-9.
59. Task Force on Abortion: 1980, *Report*, The American Lutheran Church, Minneapolis.
60. Task Force on Ethical Issues in Medicine: 1977, *Health, Life and Death: A Christian Perspective*, The American Lutheran Church, Minneapolis.
61. Thielicke, H.: 1964, *The Ethics of Sex*, trans. J. Doberstein, Baker Book House, Grand Rapids.
62. Thielicke, H.: 1966, *Theological Ethics*, Vol. 1, trans. W.H. Lazareth, Fortress Press, Philadelphia.
63. Thielicke, H.: 1970, *Death and Life*, trans. E. Schroeder, Fortress Press, Philadelphia.
64. Thielicke, H.: 1976, *The Doctor as Judge of Who Shall Live and Who Shall Die*, trans. E. Cooperrider, Fortress Press, Philadelphia.
65. Tiefel, H.: 1976, "The Cost of Fetal Research: Ethical Considerations", *New England Journal of Medicine*, 294, 85-90.
66. Tiefel, H.: 1978, "The Unborn: Human Values and Responsibilities", *The Journal of the American Medical Association* 239, 2263-2267.
67. Tiefel, H.: 1986, *The Nontreatment of Seriously Handicapped Newborns*, The Lutheran Church in America, New York.
68. Tilberg, C. (ed.): 1970, *Sex, Marriage and Family: A Contemporary Christian Perspective*, Lutheran Church in America, New York.

ROBERT L. SHELTON

RECENT DEVELOPMENTS IN MEDICAL ETHICS IN THE METHODIST TRADITION

The spiritual and ecclesiastical descendants of John Wesley have a significant history of concern for individual and social well-being and justice, including health and health care. The Oxford-based group organized by John and Charles Wesley were somewhat derisively dubbed "Methodist" for their methodical approaches to devotional piety, personal morality, and what would now be considered matters of social justice. These "fathers of Methodism...ministered to the physical, intellectual and social needs of the people to whom they preached the gospel of personal redemption" ([4], p. 3). Sometimes taking "forthright positions on controversial issues involving Christian principles", "early Methodists expressed their opposition to the slave trade, to smuggling, and to the cruel treatment of prisoners" ([27], p. 15). Even though the Methodist Episcopal Church, organized in the United States by the followers of Wesley, was to produce a number of divisions over racial, doctrinal, and ecclesial issues, the general tradition of "Methodism" remains one characterized by a combination of concerns for personal "holiness" or spiritual growth, and personal and societal welfare.

Those in the Methodist tradition have consistently shown a moral commitment to expression of beliefs based on biblical and theological understanding regarding individual and social practices. This tradition has also shown, with occasional lapses, respect for moral diversity within the religious communities and the larger political order. These trends have led to the contemporary situation in which particular positions are taken (sometimes with passionate commitment to social action) that press for wide moral consensus, even while taking care to preserve basic principles of separation of church and state.

The following discussion of developments in certain areas of medical ethics draws on statements and documents of official stature in The United Methodist Church, the Christian Methodist Episcopal Church, and the African Methodist Episcopal Zion Church. Although Methodist history also includes several other divisions, such as African Methodist Episcopal, Free Methodist, Wesleyan Methodist, and Nazarene churches, documents available at the time of this writing limit reference to the aforementioned three churches.

1. NEW REPRODUCTIVE TECHNIQUES

New reproductive techniques are not discussed in any recent official documents of churches in the Methodist tradition. The Christian Methodist Episcopal

Church urges caution that such scientific efforts as "fertilization of human eggs outside the uterus" not be misused by persons who would "exercise excessive control over" others ([5], pp. 10-11). Otherwise, it may be assumed that in vitro fertilization, artificial insemination, and other such techniques have simply not been seen as significant ethical issues in this branch of Protestantism.

Some inferences, however, may be made from positions taken on related issues. Statements whose context seem to be related mostly to contraception and abortion are open for interpretation regarding "new" techniques. Churches and "common society", for instance, are urged by The United Methodist Church to be prepared to "offer sound ethical counsel to those facing birth-planning decisions affected by...biomedical research" ([23], p. 112). Likewise, the Church is expected to "offer informed counseling and support to both men and women on all options regarding childbearing", and to be sure that its medical and mission facilities "provide a full range of fertility-related and family-planning information and services" ([22], pp. 285-286). In treatments of sexual relationships, it is clear that the only fully accepted sexual relationship is in heterosexual marriage ([27], p. 19; [4], p. 8). Thus, to the extent that any "new" reproductive method is an extension of, or substitute for, the "normal" technique of sexual intercourse, such techniques would be most fully and clearly supported only within a heterosexual marriage relationship. At the same time, United Methodist statements of recent years regarding "family" have emphasized the reality that family is experienced in "varying forms", including single parents ([27], p. 18) and "those who choose to be single" ([23], p. 112). In its resolution on "Responsible Parenthood", this denomination affirms that "[t]he family in its varying forms constitutes the primary focus of love, acceptance and nurture, bringing fulfillment to parents and child" ([23], p. 110). It may be, therefore, that any future attention to new reproductive techniques will require coming to terms with the clear affirmation that the acceptable relationship for procreation is within heterosexual marriage, whereas the recognition of "varying forms" of family as foci of love and nurture may support reproduction outside the traditional marriage relationship as long as it does not involve direct sexual activity.

2. ABORTION

The complexity of the issue of abortion is borne out in the positions taken, and the reflective statements adopted, by branches, agencies, and personnel within Methodism. The sensitivity of the problem, and the ambiguity with which it is treated, is well demonstrated in this statement adopted by the General Conference of the Christian Methodist Episcopal Church:

> Issues around the question of abortion are often very complex and always very serious. There are no simple solutions to complex matters. We believe in the sanctity of unborn human life; hence, we reject the use of abortion as a contraceptive method. The decision to have an abortion should never be a routine or casual decision. The decision to terminate a pregnancy should be made only

after all life-affirming, life-respecting alternatives have been explored with someone who is sensitive, caring and yet objective.

On the other hand, we also respect the sacredness of the life and well-being of the expectant mother and recognize the tragic and mysterious conditions of life which may make abortion an option for serious consideration. In any such occurrences, we strongly advise proper medical guidance and spiritual counsel to help avert any emotional or physical damage to the prospective mother ([7], pp. 14-15).

This recognition that abortion may be a morally acceptable option finds more support from the predominantly white United Methodists than from another American body, the African Methodist Episcopal Zion Church. The A.M.E. Zion Board of Bishops, in their Quadrennial Episcopal address to the General Conference, attacked the church's previous avoidance of the issue and declared abortion "wrong", a "sin against humanity", and thus "an abomination to God". Taking the position that life begins at fertilization, the address declares that any termination of that life from that point on is a violation of God's will. The statement does, however, recognize responsibility to minister to those who have "chosen the way which is wrong", moving to restore them to fellowship and ministering to the sufferings that accompany such a choice ([1], pp. 40-42).

The section on abortion in The United Methodist Church's *Social Principles* was amended in 1988 to include the sentence, "We cannot affirm abortion as an acceptable means of birth control, and we unconditionally reject it as a means of gender selection" ([27], p. 20). Some strong opponents of abortion in the denomination have focused on this sentence to support a claim that the church is now firmly in the anti-abortion camp. A careful and thorough response to this claim was released in 1989 by the Associate General Secretary, Health and Welfare Ministries Department, of the church's General Board of Global Ministries. Cathie Lyons introduced her discussion with these words:

We do harm to the integrity of the church's witness to complex and critical issues by focusing on the singular concern of whether the church is for or against abortion. The issues underlying the abortion debate are important to the church, and it has demonstrated caution and care with regard to life and life's potential and those situations which put the two in conflict with one another ([30], p. 1).

As Lyons demonstrated in that paper and in a subsequent briefing on the issue, The United Methodist Church has a 20-year history "of faithful witness to complex issues". A study of existing abortion laws was recommended as early as 1968 in a study document which suggested that "abortion may be justified from a Christian standpoint" under certain circumstances ([30], p. 1). The denomination's first official position on abortion appeared in a resolution on "Population Crisis", adopted by a special session of the General Conference two years later, in which it was urged that regulation of abortion be removed from the criminal code, "placing it instead under regulations relating to other procedures of standard medical practice" ([30], p. 2). This concern for safeguarding the "legal option of abortion under standards of sound medical practice" was to be included in the Resolution on Responsible Parenthood, adopted in 1976, and is still included in the (most recent) 1988 *Book of Resolutions*. The Church's broad view on the complexity of the issues is spelled out in that resolution:

When, through contraceptive or human failure, an unacceptable pregnancy occurs, we believe that a profound regard for unborn human life must be weighed alongside an equally profound regard for fully developed personhood, particularly when the physical, mental, and emotional health of the pregnant woman and her family show reason to be seriously threatened by the new life just forming. We reject the simplistic answers to the problem of abortion which, on the one hand, regard all abortions as murders, or, on the other hand, regard abortions as medical procedures without moral significance.

When an unacceptable pregnancy occurs, a family, and most of all the pregnant woman, is confronted with the need to make a difficult decision. We believe that continuance of a pregnancy which endangers the life or health of the mother, or poses other serious problems concerning the life, health, or mental capability of the child to be, is not a moral necessity. In such cases, we believe the path of mature Christian judgment may indicate the advisability of abortion. We support the legal right to abortion as established by the 1973 Supreme Court decision. We encourage women in counsel with husbands, doctors, and pastors to make their own responsible decisions concerning the personal and moral questions surrounding the issue of abortion ([23], pp. 110-111).

This resolution is consistent with that part of the "Social Principles" that immediately precedes the 1988 addition so often quoted by anti-abortionists. Reluctance to approve abortion is affirmed due to "our belief in the sanctity of unborn human life". An equal commitment, however, is given to "sacredness of the life and well-being of the mother, for whom devastating damage may result from an unacceptable pregnancy". Earlier Christian teachings are appealed to in recognizing "tragic conflicts of life with life that may justify abortion"; in such cases, "the legal option of abortion under proper medical procedures" can be supported. All Christians are called to a "searching and prayerful inquiry" into the kinds of circumstances that might make abortion acceptable. Finally, "a decision concerning abortion should be made only after thoughtful and prayerful consideration by the parties involved, with medical, pastoral and other appropriate counsel" ([27], p. 20).

Further discussion of abortion issues brings out the following:

- Neither economic status nor age should limit the availability of abortion to women ([23], p. 112).
- Minors who are pregnant should be treated as adults, and thus not be subjected by law to parental approval for treatment ([23], p. 112).
- Counseling, including religious, must be available to facilitate responsible decisions in such matters as abortion ([14], p. 239).
- The right both to "adequate and objective counseling" and to "medically accepted treatment" in regard to pregnancy and abortion is affirmed as something to be safeguarded by doctor-patient confidentiality for persons

regardless of age ([18], p. 267).

- Concerns for population include a call for the church to "exert leadership in making possible the safe and legal availability of...abortion where appropriate" ([22], p. 285).
- Concern for the status of women includes the exercise of choices concerning fertility, including safe and legal abortion, as well as being able to continue a pregnancy without threat to public financial support ([28], pp. 313-314).
- Knowledge of and access to family planning services should be aided by the church; the church, likewise, is encouraged to promote the provision of public funds for abortion services where needed in family planning programs ([28], p. 316).
- The effort to call a Constitutional Convention for the purpose of declaring the fetus a person from the moment of conception is strongly opposed; such an effort is considered to be contrary to the principle of separation of church and state, because it would write one particular theological viewpoint into the U.S. Constitution. The resolution notes that the theological perspective from which the call for a Constitutional Convention is made is not shared by a number of religious bodies, including The United Methodist Church ([20], p. 461).

3. MATERNAL-FETAL CONFLICT

Issues regarding the tragic conflict of life with life are addressed in the above discussion on abortion. The A.M.E. Zion Church noted that the debate has included concern for the mother on the one hand, and the fetus on the other, but does not specifically address the inherent conflict while concluding that abortion is categorically "wrong" ([1], pp. 41-42). The Christian Methodist Episcopal Church statement emphasizes the complexity of the issues, affirming the "sanctity of unborn human life" while respecting the "sacredness of the life and well-being of an expectant mother" ([7], pp. 14-15).

The United Methodist Church has noted a historic recognition of the "tragic conflict of life with life" ([27], p. 20), leading to a contemporary "profound regard for unborn human life" that is to be "weighed alongside an equally profound regard for the fully developed personhood" of the mother ([23], pp. 110-112). This church also emphasizes that a range of other considerations may be taken into account in that conflict, including the "physical, mental and emotional health" of both the woman and her family, as well as "serious problems concerning the life, health or mental capability of the child to be". When such conflicts exist, they concluded, continuance of such a pregnancy is "not a moral necessity", and "the path of mature Christian judgment may indicate the advisability of abortion" ([23], pp. 110-111). Such decisions should always include counsel with spouse, doctor, and pastor, to the extent they are available and appropriate.

4. CARE OF SEVERELY HANDICAPPED NEWBORNS

"Every child...has the right to be regarded as a person and shall have the right to receive appropriate medical care and treatment" ([18], p. 266). This affirmation in a United Methodist document, followed by the assertion that "[e]very person, regardless of age, shall have the right to considered and respectful care" ([18], p. 268), certainly suggests a basis for supporting care of severely handicapped newborns. Another resolution calls for resources through the church to be provided "for persons with various handicapping conditions". The goal of such resource provision is to guarantee that "each individual has full opportunity for growth and self-realization with the community of faith and the society at large" ([26], p. 281). One may very well ask, then, whether a severely handicapped newborn will actually have the capacity for "growth and realization" in its relationship with the larger community, and to what extent the community is obliged to protect and guarantee the right to care. At this point, no Methodist denomination has given policy direction in this issue. All indications, however, would point to an expectation that "considered and respectful care" would be provided to a handicapped newborn, irrespective of the severity of the condition. The nature and limitations of such care are explored in the case study and discussion section of the United Methodist-Roman Catholic dialogue report ([9], pp. 19-20).

5. CONSENT TO TREATMENT AND EXPERIMENTATION

United Methodists insist that "those engaged in research shall use human beings only after obtaining full, rational and uncoerced consent" ([27], p. 25). Likewise, the necessity for adequate information and counseling is affirmed in the face of "difficult medical choices", including experimentation, institutionalization, and various forms of treatment, as well as termination of treatment ([27], p. 21; [14], p. 239; [9], pp. 9, 12). Self-determination and the exercise of autonomy are required for those made in the image of God; thus, the patient must be "an active participant in medical treatment decisions" ([9], p. 12). This includes encouragement for "advance directives" which provide for "proxy" or "power of attorney" arrangements to protect one's desires and consent.

6. CONFIDENTIALITY

The principle of confidentiality in health care has been thoroughly affirmed in resolutions adopted by The United Methodist Church. The right to doctor-patient confidentiality is particularly extended to contraception, treatment for venereal disease, counseling and treatment related to abortion, psychiatric care and counseling, and drug or alcohol dependency ([18], pp. 266-267). Likewise, voluntary testing for and counseling related to AIDS must be protected by confidentiality ([11], p. 104).

7. EQUITABLE ACCESS TO HEALTH CARE

In its "Social Creed" the Christian Methodist Episcopal Church stands for "the provision of adequate medical care for all people", and would give special attention to "the aging, the young, and low-income individuals and groups" ([4], p. 5). The United Methodist Church has several references to a "just society" and a "just health system" in officially adopted resolutions. These statements flow from an affirmation that each person in a just society has a *right* to "basic health services that are accessible and affordable in each geographic and cultural setting" ([14], p. 237). This "right" is bluntly stated in the church's "Social Principles", with the right to health care as one of a list of freedoms that governments are expected to guarantee ([27], p. 30). Public financing is encouraged as a part of an "overall health insurance plan" to assure that persons who cannot afford premium payments are covered ([27], p. 30; [15], p. 245). Concern is expressed that health care be available and accessible that is "sensitive to cultural needs", with "compassionate and skilled care" ([14], p. 237).

No one single system can meet the needs of all communities, and United Methodists are urged to participate in decisions at a level that assures particularities of service meeting unique kinds of needs: "The system should display the requisite diversity and adaptability to meet differing social and community circumstances" ([15], p. 243). This denomination, from its origins in eighteenth-century England, has a strong history of direct involvement in health care, and its official statements affirm a continuing commitment to the church's active role in the provision of health care service. Thus, although it is not assumed that government can do everything, government is asserted to have an active role at all levels, and is expected to perform its responsibility as guarantor of rights to medical and health care.

The principles and goals identified by this denomination are placed in a context of holistic care, with the spiritual dimensions of health and health care usually central to its official statements. Thus, individual United Methodists are urged to "[r]ecognize the need for certain fundamental changes in the way health care services in the U.S. are presently organized, delivered and financed", as well as to "[c]hoose those individual and family life styles which are conducive to good health" ([15], p. 246).

The United Methodist Church, as an international network of churches and ministries, is also committed to support of the World Health Organization's "Health for All by the Year 2000". Articulation of this commitment to global access to health care specifically recognizes, as have other health care statements, the complex set of dimensions to be taken into account, with individual and family values and behavior interacting with social, economic, cultural, and political conditions ([16], p. 247).

Other documents focused on specific issues repeat this right to access to care. Addressing "Medical Rights for Children and Youth", the right of "every person, regardless of age", to "timely access to continuing and competent health care" is given the particularly strong status of a "canon" ([18], p. 266). Included in this right are basic, emergency, and psychiatric medical care. Also emphasized are both information and treatment, on a confidential basis, relating to contraception and birth-control, venereal disease, pregnancy and abortion, and drug or alcohol dependency. Of particular interest is insistence on the protection of the "privacy of every patient, regardless of age"; if the patient desires, observers are to be removed from the examining area ([18], p. 268). This document takes a firm position supporting community intervention in behalf of children when parents or legal guardians refuse consent to the care, suggesting somewhat detailed procedures to be followed in such eventualities ([18], pp. 267-268). The "Responsible Parenthood" resolution urges "to each pregnant woman accessibility to comprehensive health care and nutrition adequate to assure healthy children" ([23]. p. 111). This document reiterates the above described concerns regarding sexual and family planning services for minors, as well as for adults. It also supports access to both voluntary sterilization and abortion "without regard to economic status" ([23], pp. 111-112).

With regard to care for the dying, United Methodist partners in dialogue with Roman Catholics propose the creation of a context that assures "equitable access for all to the resources that will relieve the dying and their loved ones of anxieties about financial crises created by medical expenses, while providing holistic care to accompany patients through their dying" ([9], p. 15]).

United Methodist concerns for access are summarized in this way:

> In a just society all people are entitled to basic maintenance and health care services. We reject as contrary to our understanding of the gospel the notion of differing standards of health care for various segments of the population....Health care should be comprehensive, including preventive, therapeutic, and rehabilitative services ([14], pp. 238-239).

8. ETHICAL ISSUES POSED BY COST-CONTAINMENT MEASURES

As a part of the church's "health and wholeness" ministry, United Methodists are challenged to support "policies and programs that will ensure comprehensive health care services of high quality" to all. In the United States, it is recognized that attempts to provide such services with equal access leads to issues of cost-containment, and the church affirms the findings of the President's Committee on Medical Ethics in 1983, quoting from that report: "Measures designed to contain health care costs that exacerbate existing inadequacies or impede the achievement of equity are unacceptable from a moral standpoint" ([14], p. 241).

In addressing some of the social dimensions of ministry to persons who are dying, the United Methodist team in dialogue with Roman Catholics noted certain "failures in distributive justice" in relation to cost containment. The term "cost containment" is not specifically used in the dialogue report. It does, however, recognize that both private and government health care plans, in response to allocation issues, may fund certain diagnoses and treatments that are technologically sophisticated, "while denying or minimizing payments for less costly services...critical for humane and dignified dying". Likewise, it is recognized that payment policies dictating treatment length and modalities often produce constraint on health care professionals "in their efforts to implement care plans that have patient benefit as their goal". Reversal of such policies is urged, with higher priority given to hospice, home care, and social services ([9], p. 14).

9. WITHDRAWING/WITHHOLDING CARE

In the "Social Principles" of The United Methodist Church, the denomination asserts "the right of every person to die in dignity, with loving personal care and without efforts to prolong terminal illnesses merely because the technology is available to do so" ([27], p. 21). Concern for "extreme measures to prolong life" likewise led to support of appropriate counseling to be made "available to all patients and families" when confronted with difficult medical choices of this sort ([14], p. 239).

With this foundation, United Methodist scholars in dialogue with Roman Catholics were secure in affirming the development of a context in which persons can "accept relief of suffering as a goal for care of the dying rather than focusing primarily on cure or prolongation of life" ([9], p. 15). This affirmation was accompanied by a recognition of some issues in the "ethos of the medical profession":

> The emphasis on curing, healing and restoration contributes to uneasiness among physicians in making the transition from cure to care when the possibilities of cure are exhausted. Members of the medical profession must accept the legitimacy of medicine oriented toward relief of suffering rather than extension of life for the dying, and they must use their skill in identifying when it is appropriate to discontinue treatment. This is not easily done: there is often a continuum of treatment for life-threatening conditions and a tendency to seek one more procedure that can legitimately prolong a patient's life ([9], p. 14).

These concerns are echoed throughout the dialogue document. Dying individuals have the right to be protected from "the extremes of premature or delayed termination of treatment" ([9], p. 12). (This statement, of course, reflects a similar caution about euthanasia, to be discussed in the next section.) A provocative affirmation in the opening theological and ethical section of the document sets the stage for further ethical and pastoral care debate in the church bodies:

> We affirm that the obligation to employ life-sustaining treatments ceases when the burdens (physical, emotional, financial, or social) for the patient and the caregivers exceed the benefits to the patient. The application of excessive procedures, sometimes encouraged by the ingenuity of modern medical technology, does not reflect good stewardship because it does not serve the purpose for which God gave life ([9], p. 6).

The ethical methodology for "determining what constitutes Christian care" is outlined: "we should engage our biblical and doctrinal understanding with the wisdom of new knowledge about human life and with our own experience of the reality of disease and death" ([9], p. 6).

The United Methodist and Roman Catholic dialogue teams further agreed that a measure of maintaining dignity in dying is participation in "cognitive and affective activities that enable conscious, loving relationships with others in community". The gift of medical technology supports the preservation of such dignity "to the extent that it is used to sustain, support, and compensate for human functions". A crucial warning follows:

> However, when technology becomes an end in itself, unduly prolonging the dying process, it creates a paradox in which human dignity may be undermined and where the goals of treatment are distorted to accommodate the imperatives of technology. When a person is dying and medical intervention can at best prolong a minimal level of life at great cost to human dignity, structures of care and use of medical technology should focus on maximizing the individual's capacity for awareness, feeling, and relationships with family and community. Decisions that subordinate the humane dying of a terminally ill man or woman to the technological imperative, or personal or institutional self-interest – legal, financial, professional – are not consistent with Christian values and traditions ([9], p. 11).

The exercise of self-determination in termination of treatment is likewise urged in the document, through advance directives (i.e., "living wills") and arrangements for durable power of attorney. Significant emphasis is also placed on hospice and home care treatments that emphasize "care", rather than on the "technological imperative" related to an emphasis on "cure" that is no longer realistic in particular cases. This discussion includes agreement with a position taken in the Vatican Declaration on Euthanasia in 1980, that pain control methods whose intention is to relieve pain rather than to kill, may be "used for terminally ill patients...even when they risk or shorten life" ([9], p. 13).

This openness to termination of treatment in some cases led the General Board of Church and Society of The United Methodist Church to join in a brief of *amicus curiae* supporting the petition of Nancy Beth Cruzan's parents to the U.S. Supreme Court to allow termination of nutrition and hydration. In the brief, it is argued that a traditional moral consensus, a widely held tradition of conscience in the United States, supports a person's right to "decline life-prolonging treatment". Although it is not assumed, or even urged, that this group's religious perspectives are necessarily to be equated with correct constitutional interpretation, it is asserted that society's tradition of conscience has been informed by principles held by this and other religious organizations.

These principles include: (1) "[L]ife should not be assessed in purely medical terms. Life and health...are an integration of the spiritual, emotional and physical aspects of being"; (2) "[P]rolonging the physical life of the body at all costs and in whatever form is radically inconsistent with...proper respect for life, and for death"; and (3) The family is the proper context for decisionmaking regarding "how best to cope with the natural ending of a life", and government should not intrude in even a surrogate role of the family ([32], pp. 5-6). Endorsement of active euthanasia, or "the *premature* termination of human life to avoid suffering", is not intended by the writers of this brief ([32], p. 10). A central ethical declaration of the brief is posed in these words:

A person choosing to meet death naturally in the event of a permanent loss of all conscious existence is in no sense making an aberrant or eccentric choice. To the contrary, it is a choice fully consonant with traditional understandings of the proper bounds of life and the place of death. Slavish application of technology to preserve elemental physical functioning, in contrast, may well degrade the life, spirit and memory of the patient by preventing both patient and family from making peace with the end of life ([32], p. 10]).

Sources utilized in the brief include not only the relevant resolutions and other United Methodist documents, but also reports from agencies of the Presbyterian Church, U.S.A., the report of the President's Commission for the Study of Ethical Problems in Medicine and Biomedical Behavioral Research, and previous Supreme Court decisions ([32]).

10. ACTIVE EUTHANASIA

The Christian Methodist Episcopal Church, a predominantly African-American denomination, is clear in its opposition to active euthanasia: "The Christian Methodist Episcopal Church takes the position that putting persons who are suffering from incurable diseases and unbearable pain to death 'mercifully' places some human beings in the 'God-role' of deciding who has the right to life and who should be put to death. Such a decision is inconsistent with our belief in God's will to restore, heal and recreate the brokenness of His world, even in the face of death" ([6], p. 14]).

United Methodists, on the other hand, have not stated a position on this matter in any official form. The more "passive" matter of allowing persons to die by termination or withdrawal of treatment is supported in a number of ways, as demonstrated in Section 9. Such statements, however, sometimes include a "disclaimer" that distinguishes allowing to die from active euthanasia. As previously indicated, the brief in support of the *Cruzan* appeal to the U.S. Supreme Court stipulated that its arguments were in no way an endorsement of active euthanasia. United Methodists in dialogue with Roman Catholics agreed that "the direct intentional termination of innocent human life, either of oneself or another, has been generally treated in Christian tradition as contradictory to [loving stewardship of life] because it is a claim to absolute dominion over human life" ([9], p. 6). They also affirmed that "withholding or withdrawal of medical interventions of little benefit to the patient compared with the burdens they impose should not be confused with euthanasia or suicide" ([9], p. 12). Dying patients have the right to be protected from "premature...termination of treatment", just as they are to be protected from the delay of such termination ([9], p. 12).

It is also noted in the dialogue report that some of the United Methodist participants recognized there may be exceptions to "unequivocal rejection of suicide and active euthanasia", and that "euthanasia...might be an ethically permissible action" in some circumstances ([9], pp. 6, 13). The taking of one's own life in the face of "terminal or irreversible medical conditions" ([9], p. 13) is beginning to be considered as a possible exception to the long tradition of absolute prohibition of suicide. Pastoral caregivers are urged to enter into discussion with persons contemplating such action, as well as with their family members and other loved ones ([9], p. 9). Such communication would assist "in understanding God's gift of life, the human stewardship of life, and the saving action of Christ in his suffering, death and resurrection". In addition, "the loving presence of Christ as manifested in the church community should surround those contemplating suicide and the survivors of those patients who take their own lives" ([9], p. 9).

Some recent developments within The United Methodist Church in regard to suicide are of particular interest. A resolution adopted in the most recent General Conference of the church was developed in response to the troubling phenomenon of significant increases in teen-age and young adult suicides. The resolution, however, while taking account of important demographic, societal, and causal issues, insists at the outset that "[a] Christian perspective on suicide...begins with an affirmation of faith: suicide does not separate us from the love of God" ([25], p. 317). "Major initiatives" by the church are encouraged for the prevention of suicide, with the recognition that a society that lacks a "caring attitude toward all persons" will experience an increase in suicide. The resolution does not specifically address the hastening of death by persons with terminal medical conditions. It does assert the traditional faith perspective that bodily death is not the end of life nor of one's relationship with God, even though it is a loss, often severe, in community ([25], p. 322).

Closely related to the development of this resolution was the publication of an "Occasional Paper" on "Suicide, The Bible, and Ethics in Contemporary American Society", authored by Dr. James T. Clemons of Wesley Theological Seminary, and published by The United Methodist Board of Higher Education and Ministry. Clemons noted the centuries-long Christian position that suicide is a sin (and a crime) because it is a human act to end life that began as a gift of God. He also discussed developments since the Enlightenment that open the question of human use of God-given rights to choose the ways in which one's life will end. As a biblical scholar, he is interested in the fact that the Hebrew scriptures (*Old Testament*) relate six direct accounts of suicide, and the

Christian *New Testament* includes one direct account ([8], pp. 6-8). All of these are simply reported in the biblical text, with no judgment suggested. Some were honored in their deaths, while others were given normal burials. Indirect texts that have been traditionally utilized by Christians as prohibitions of suicide ([8], p. 8) are discussed by Clemons, as well as several texts that support the Christian's "laying down his or her life and so fulfilling the law" ([8], pp. 8-9). After outlining some theological and ethical issues that need to be pursued, and asserting that only in the area of pastoral care is the church responding to the contemporary suicide crisis, Clemons concludes that, "[u]ntil those in the academic fields of Bible and ethics take their several tasks regarding suicide seriously, we shall, by default, abandon the field to an uninformed, unenlightened, and medievally-oriented perspective" ([8], p. 9).

The suicide issue was also addressed in a significant and unique manner through the "Consultation on 'Self-Deliverance'" held at the Iliff Institute in September 1989. This Institute, related to a United Methodist seminary, Iliff School of Theology, brought together a group of scholars, pastoral care specialists, and persons deeply involved in ministry with persons suffering from AIDS or AIDS Related Complex. As was stated by the Director of the Colorado AIDS Project (a United Methodist minister) in a letter to participants in the consultation, "the desire to end one's life in peaceful dignity has emerged as a viable option for a growing number of people with AIDS", faced with the devastation of withering and wasting away "in agonizing death" [29]. The consultation was an attempt to understand those who held such views, as well as to examine closely the issues faced by those who were attempting to provide medical, social and pastoral care. The concept of "self-deliverance" was proposed as a way of thinking theologically and ethically about a faithful act of delivering oneself from life into death within the context of a faith that affirms eternal life in God's grace [29].

11. DEFINITION OF DEATH

A definition of death is not found in any documents adopted by Methodist official bodies. It is worth noting, however, that the General Board of Church and Society of The United Methodist Church turned to a Presbyterian statement defining death in its *amicus curiae* brief to the United States Supreme Court in the *Cruzan* case:

> [W]hen the capacity for human relationship is irretrievably lost, death has occurred regardless of what biological function can be sustained. In medical terms, that means when brain function ceases and when a flat electroencephalogram occurs, cardiovascular activity ceases, or other tests of responsiveness have been conducted and found to be negative. It is then that we recognize that death is in the midst of life, even as we celebrate the life that overpowers death ([32], pp. 9-10).

Some might find this statement in need of qualification and some elaboration. At the same time, nothing in official United Methodist documents

– or any other available, recent statements from others in the Methodist tradition – would suggest that the General Board of Church and Society was out of line in utilizing this statement in their brief.

12. ORGAN DONATION AND TRANSPLANTATION

The United Methodist Church firmly supports the donation of body organs and tissues for transplantation. Proceeding from the affirmation that "selfless consideration for the health and welfare of others is at the heart of the Christian ethic", such an act is particularly recognized as "life-giving". It is also affirmed as providing a source of comfort to surviving loved ones, especially when it may be "perceived as a positive outcome of a seemingly senseless death" ([21], p. 108). Because of the world-wide recognition by "moral leaders" of this humanitarian opportunity, and because many die who could otherwise live with such aid, members of the denomination are urged to "become organ and tissue donors by signing and carrying cards or driver's licenses, attesting to their commitment of such organs upon their death, to those in need" ([21, p. 108). Furthermore, those who provide counseling to patients and families facing difficult medical decisions are encouraged to assist in "responsible decisions, within the context of the Christian faith" regarding a number of matters, including organ transplants ([14], p. 239). United Methodist representatives were thus well grounded in their own community's commitments when they agreed with their Roman Catholic dialogue partners that "[t]he gift of life in organ donation allows survivors to experience positive meaning in the midst of their grief and is an important expression of love in community". These practices are supported by both churches "as long as death is not hastened and is determined by reliable criteria". Specific options should be explored by pastoral caregivers "with patients and their families" ([9], p. 9).

The University of Kansas
Lawrence, Kansas, U.S.A.

NOTE

The United Methodist Church has adopted an official policy that "The" is to be used, and capitalized, in any reference to the church. The author has attempted to follow that policy throughout this draft.

BIBLIOGRAPHY

1. Board of Bishops of the African Methodist Episcopal Zion Church: 1984,

The Quadrennial Episcopal Address, 42nd Session of the General Conference, St. Louis, Mo., July 25 - August 3, 1984, African Methodist Episcopal Zion Church, Charlotte, North Carolina.

2. Board of Bishops of the African Methodist Episcopal Zion Church: 1988, *The Quadrennial Episcopal Address, 43rd Session of the General Conference, Charlotte, N.C.*, African Methodist Episcopal Zion Church, Charlotte, North Carolina.
3. Board of Social Concerns, Christian Methodist Episcopal Church: n.d., *Denominational Statements on Issues of Social Concern*. Christian Methodist Episcopal Church, Atlanta.
4. General Conference, Christian Methodist Episcopal Church: 1966, "Social Creed", in [3], pp. 3-8.
5. General Conference, Christian Methodist Episcopal Church: 1978, "Scientific Research", in [3], pp. 10-11.
6. General Conference, Christian Methodist Episcopal Church: 1978, "Euthanasia", in [3], p. 14.
7. General Conference, Christian Methodist Episcopal Church: 1978, "Abortion", in [3], pp. 14-15.
8. Clemons, J.: 1987, "Suicide, The Bible and Ethics in Contemporary Society", *Occasional Papers* 72, The United Methodist Board of Higher Education and Ministry, Nashville.
9. Delaney, J., and B. Oliphint (eds.): 1989, *Holy Living and Holy Dying - A United Methodist/Roman Catholic Common Statement, Prepared by the General Commission on Christian Unity and Interreligious Concerns of The United Methodist Church and the Bishops' Committee for Ecumenical and Interfaith Affairs*, National Conference of Catholic Bishops, General Board of Global Ministries, Cincinnati.
10. The General Conference of The United Methodist Church: 1988, *The Book of Resolutions of The United Methodist Church - 1988*, The United Methodist Publishing House, Nashville.
11. The General Conference of The United Methodist Church: 1988, "AIDS and the Healing Ministry of the Church", in [10], pp. 101-106.
12. The General Conference of The United Methodist Church: 1988, "Federal Funds for Indian Health Services", in [10], pp. 211-212.
13. The General Conference of The United Methodist Church: 1988, "Genetic Science", in [10], pp. 213-216.
14. The General Conference of The United Methodist Church: 1984, "Health and Wholeness", in [10], pp. 235-241.
15. The General Conference of The United Methodist Church: 1980, "Health Care Delivery Policy Statement", in [10], pp. 241-247.
16. The General Conference of The United Methodist Church: 1984, "Health For All By the Year 2000", in [10], pp. 247-249.
17. The General Conference of The United Methodist Church: 1980, "Human Rights", in [10], pp. 445-447.

18. The General Conference of The United Methodist Church: 1976, "Medical Rights for Children and Youth", in [10], pp. 265-268.
19. The General Conference of The United Methodist Church: 1988, "Mental Health", in [10], pp. 268-272.
20. The General Conference of The United Methodist Church: 1980, "Opposition to a Call for a Constitutional Convention", in [10], pp. 458-462.
21. The General Conference of The United Methodist Church: 1984, "Organ and Tissue Donation", in [10], p. 108.
22. The General Conference of The United Methodist Church: 1980, "Population", in [10], pp. 283-287.
23. The General Conference of The United Methodist Church: 1976, "Responsible Parenthood", in [10], pp. 110-112.
24. The General Conference of The United Methodist Church: 1980, "Safety and Health in Workplace and Community", in [10], pp. 370-373.
25. The General Conference of The United Methodist Church: 1988, "Suicide: A Challenge to Ministry", in [10], pp. 317-322.
26. The General Conference of The United Methodist Church: 1984, "The Church and Persons With Mentally, Physically and/or Psychologically Handicapping Conditions", in [10], pp. 279-283.
27. The General Conference of The United Methodist Church: 1988, "The Social Principles", in [10], pp. 15-36.
28. The General Conference of The United Methodist Church: 1980, "The Status of Women", in [10], pp. 311-316.
29. Iliff Institute: 1989, *Reflections on Self-Deliverance*, unpublished study papers for The Consultation on Self-Deliverance, Iliff Institute, Denver.
30. Lyons, C.: 1989, *Abortion and Responsible Parenthood: The Faithful Witness of a Church to Complex and Critical Issues*, Health and Welfare Ministries Department, General Board of Global Ministries, The United Methodist Church, New York.
31. Lyons, C.: 1989, *On Not Misrepresenting The Faithful Witness of The Church*, Health and Welfare Ministries Department, General Board of Global Ministries, The United Methodist Church, New York.
32. Martin, T., D. Verelli, and M. Lopez: 1989, *Brief of the General Board of Church and Society of The United Methodist Church As Amicus Curiae in Support of Petitioners, Nancy Beth Cruzan, by Her Parents and Co-Guardians, Lester L. and Joyce Cruzan, in the Supreme Court of the United States, October, 1989* (includes references to: "Life Abundant: Values, Choices and Health Care, The Responsibility of The Presbyterian Church, U.S.A.", a Report of the Advisory Council on Church and Society to the 200th General Assembly, 1988, and to "The Covenant of Life and the Caring Community", Policy Statement of the 195th General Assembly of the Presbyterian Church, U.S.A., 1983).

PAUL D. SIMMONS

BAPTIST-EVANGELICAL BIOMEDICAL ETHICS

Treating Baptists and evangelicals in the same article is both problematic and understandable ([24]; [48]). Historically, Baptists (American and Southern) are defined by commitments to religious liberty and local church autonomy. They have been wary of creeds, because of the tendency to displace faith with doctrinal beliefs. Evangelicalism is a difficult term to define, and clarifying who is and who is not an evangelical is even more troublesome. Conferences have been convened to attempt to discover common affinities among diverse groups who claim the designation [21].

General definitions typically range over those Protestant Christians who believe in biblical authority, a personal experience of salvation, evangelism,and congregational polity. More restrictive views want to make doctrine the central and defining concern. Typically, biblical inerrancy, virgin birth, substitutionary atonement, the bodily resurrection of Jesus, and the bodily return of Jesus form the core of beliefs necessary to being Christian. Evangelicalism would thus include fundamentalism, which is problematic, since fundamentalists are known as obscurantist and non-involved in social or political issues [22]. The term neo-evangelical was coined to distinguish between those fundamentalists who are interested in social issues and those who still define themselves as anti-modernist [48].

The media image of Southern Baptists is one much more nearly defined by fundamentalism (evangelicalism). Certainly, the leadership of the Southern Baptist Convention since 1979 has identified itself in these terms.

The question of definitions and distinctions cannot here be resolved. The purpose is to indicate responses to bioethical issues that can be discerned among those who identify themselves as evangelical or Baptist. There is no clear line of demarcation between the two, nor are there definitive positions for either Baptists or evangelicals. The reasons for various positions taken by writers, as well as their disagreements with one another, will be noted. Strong affinities and common commitments can certainly be identified, which constitute basic assumptions or points of reference in dealing with bioethics.

1. *Biblical Authority*

Evangelicals center moral concerns in biblical revelation. Attitudes toward the Bible range from notions of "inerrancy and infallibility" to concern for being consistent with the perspectives and concerns of the biblical witness. "Biblical authority" is the term that pulls them all together. The Bible is the center of gravity for dealing with questions of theology and morality, since there is a distinct and inescapable relationship between the contents of Scripture and God's will for human life and thought.

2. *Moral Fervor*

Evangelicals are serious about the moral life and tend to measure sincerity of faith by certain moral expectations. In spite of a "grace" theology, it often sounds like works righteousness. Morals are also related to law or public policy: what is regarded as immoral should be made illegal. Both Baptists and evangelicals share a Puritan past. Baptists became champions of religious liberty while evangelicals maintained a theocratic approach to law. Some Baptists still maintain that First Amendment rights are important when shaping public policy for moral issues [59].

Evangelicalism has been occupied more by theology than ethics, which has resulted in a lack of sophistication in moral reasoning. Charges of legalism are both understandable and deserved. Moral postures (as in bioethics) take on an "either-or" flavor, in which one side is identified as "Christian" and the other as "unchristian and non-scriptural" at best, and "pagan" or "secular humanist" at worst [18]. No argument is permitted, and no explanation is required. As one writer says, right actions require no moral justification. Only those contrary to God's will involve rational reflection [60].

3. *Divine Sovereignty*

A sense of the presence of the spirit of God and a belief that God is sovereign over human affairs is a strong motif in evangelicalism. God is often thought of as watchdog or policeman, an enforcer of good morals, often by punishing the erring. Individualism in private life is complemented by a stress on corporate responsibility under divine judgment [9]. Old Testament images of God often dominate such theology, contrary to preachments about New Testament Christianity. Those writers who struggle with such questions as the nature of God, historical relativism, and theodicy likely will not reflect traditional perspectives.

An analysis of responses to biomedical issues among Baptists and evangelicals will reveal patterns of thought and argument that reflect the above factors, which provide points of reference in considering the primary variables at work in bioethics.

NEW REPRODUCTIVE TECHNIQUES

New reproductive techniques are of interest to evangelicals and Baptists because of their commitments to family values. Most forms of contraceptives are accepted, since the control of fertility is thought to be morally responsible. Unfortunately, major breakthroughs such as RU486 get tangled in the abortion debate [28]. Anti-choice evangelicals have actively worked against its introduction into the United States. Moral Majority and Operation Rescue leaders claim that Hoechst AG, the German corporation that owns majority stock in the drug company that manufactured RU486, has made three commitments in writing that the drug would not be marketed outside of France. They have threatened a boycott of Hoechst AG, which does billions of dollars of business in the United States, if that "promise" is broken ([29], p. 53).

Approaches to artificial insemination (AIH, AID), surrogate motherhood, in vitro fertilization, and sex selection show major differences among writers. Concerns for the divine purpose for human sexuality and the institution of marriage permeate all writings, but conclusions and applications vary widely.

John Davis says that AIH is acceptable, since it furthers the bonds of marriage ([6], p. 72). AID is immoral, since it involves the intrusion of a third party into the marriage relationship, and it reminds the husband of his failure even though it serves to fulfill the maternal functions of the wife. Davis fears that the "third party" factor might harm the marital relationship and introduce deception about the child's true origins ([6], p. 73).

Davis opposes surrogate motherhood as the mirror image of AID, implies it is a type of prostitution, and says it is often done by the surrogate to atone for an earlier abortion ([6], p. 77). He denies the analogy to adoption, and says that the practice is clearly immoral.

Sex selection as grounds for abortion is "not an option for the Christian" ([6], p. 80). Davis believes parents should be prepared to accept in love the birth of a child not of the desired sex. He does not deal with the tough cases of sex selection because of radical genetic deficits, e.g., hemophilia, muscular dystrophy, and cystic fibrosis, which are linked to males. Ironically, his concern with abnormality in offspring is the dominant moral norm he uses in treating the issue of in vitro fertilization ([6], p. 89).

Not all agree with such negativism ([4], Ch. 7; [53], Ch. 5). There are at least seven means of attaining pregnancy that involve biotechnical assistance or intervention. Objections based on the immorality of masturbation, commercialization, unnaturalness, the demystification of sex, and concerns about abortion seem misplaced and inadequate. The moral questions should be set in the context of caring for the well-being of the intended offspring, the covenantal nature of marriage, and the importance of parenthood to the couple. Surrogacy can be an enormous gift and clearly altruistic, though caution and prudence are necessary. Bypassing infertility can be an occasion for celebrating God's blessed gift of a child. Bearing children is related to calling, not simply to biological function, and the child symbolizes the care and commitments that are required of parents in the biblical story ([53], p. 192).

ABORTION

Abortion is *the* issue that has most energized and politicized as well as deeply divided conservative Protestant Christianity ([65]; [14]). It has generated more response among the groups under discussion than all other bioethical issues combined.

During the 1980s there was a proliferation of Political Action Groups (PACs) led by evangelical religious leaders and dedicated to what they regard as a Christian agenda for the nation. Shrill moral rhetoric and a non-negotiable spirit characterized their approach to abortion both as a moral issue and as public policy. (Other issues included opposition to Gay Rights, the Equal Rights Amendment, sex education in public schools, and pornography.)

The most prominent group was the Moral Majority, led by Jerry Falwell. It is now defunct, though Falwell is still an influential voice. Other groups, such as The American Coalition for Traditional Values, led by Tim LaHaye, Billy Graham's Christian Action Council, James Dobson's Focus on the Family, and Pat Robertson's 700 Club, are still politically active against legal abortion. They combine evangelical moral and religious fervor with computerized mailing lists, well-funded campaigns, sophisticated political strategies, effective organizational efforts, and influential television programming.

The "left wing" of evangelicalism includes Evangelicals for Social Action and *Sojourners* [67] magazine, which attempt to broaden conservative Christians' interest in social issues to include such topics as capital punishment, nuclear war, and poverty. Some embrace the "seamless garment" argument in extending the notion of the sanctity of life to include human life in its social context [52]. They attempt to overcome the inconsistency of claiming to believe in the sanctity of life but applying it only to abortion and euthanasia ([17], p. 154).

Operation Rescue, founded by Randall Terry on the principle of non-violent resistance, is the most controversial approach to abortion among evangelicals. These "shock troops" of the anti-abortion movement attempt to block women's access to abortion clinics. They link arms, verbally harass women and their escorts, fall limp when arrested, and refuse to pay fines. The courts are increasingly less tolerant of their disruptive activities. Terry is now serving a two-year sentence for his activities, and heavy fines have been levied against Operation Rescue, which has declared bankruptcy. Most influential evangelicals have distanced themselves from Operation Rescue [16], but it continues to channel the energies of a radical minority, which prides itself on images of martyrdom in opposing what it calls "Nazi" and "demonic" judicial forces in America ([15], p.49; [25]).

American Baptists, in a statement from the General Board in 1988, acknowledged their "diversity of opinion" on abortion, opposed it as "a primary means of birth control", condemned the violence and harassment aimed at clinics and clients, and affirmed those who choose abortion "under difficult circumstances" and after prayerful and conscientious consideration.

Southern Baptists are deeply divided about the morality of elective abortion, as well as about the appropriate responses for molding public policy ([30];

[56]). While polls show a moderate position toward the issue among Baptists, present Convention leadership is committed to a radical anti-abortion stand. During the 1970s, resolutions passed by the annual convention called for legislation that allowed therapeutic abortions and rejected efforts to support a Constitutional ban ([2], 1971, 1974, 1977).

The ascension to power of a fundamentalist, New Right faction shifted the emphasis. In 1980 the SBC called for "legislation and/or a constitutional amendment prohibiting abortion except to save the life of the mother" ([2], 1980, p. 48). Meeting in Dallas in 1985, the Convention voted to observe "Sanctity of Life" Sunday in commemoration of the January 1973 Supreme Court decision. In 1990 a resolution was passed supporting boycotts of corporations and businesses that contribute to "pro-abortion" organizations ([2], 1990). Trustees of the Sunday School Board have mandated that all writings embrace "sanctity of life", though it is to be applied only to abortion, not to related issues such as animal rights, elective death, war, or capital punishment. The Sunday School lessons are to argue that the Bible teaches a strict anti-abortion ethic, and that Christians are to seek laws to restrict or prohibit its access.

The Christian Life Commission of the SBC has been required by its Board to revise all literature in keeping with the anti-abortion agenda [33]. The mandate includes instructions to revise statements about women, making it clear that women are not candidates for ordination. The linkage of abortion to women's issues is important to note, since women's rights do not have high priority among evangelicals. The abortion debate is set, not in terms of a conflict of rights between women and the unborn, but in terms of the superior rights of the fetus. The question of the woman's bodily integrity is of little concern when weighed against the "innocent" child and its "right to life". The issue of women's rights is simply evaded. Whether the fetus is a person in the same sense as the woman is a major variable in the disagreement about abortion [34].

The CLC has also opened a Washington, D.C., office through which it will lobby for legislation, often directly counter to the Baptist Joint Committee on Public Affairs, a church-state watchdog agency representing eight Baptist denominations. Recently, the CLC filed a friend-of-the-court brief to a challenge to federal funding of family planning clinics before the Supreme Court. The aim is to block any mention of abortion in counseling clients of such clinics [66].

Pamphlets published by the CLC prior to the current leadership recognized the silence of the Bible, the tragic circumstances of many pregnancies, the complexity of the woman's decision, and the problematic of an absolutist ban by law. Moderation on the abortion issue is no longer permitted by the agencies and commissions of the SBC.

Even so, a majority of Baptists remain committed to legislation that will permit abortion under some circumstances. Some are solidly pro-choice, believing that any legislative restriction in early pregnancy is unwarranted [53]. The fervor and intransigence on the issue of abortion within Convention leadership reflects the influence of the New Religious and Political Right. Diversity is neither admitted nor tolerated, and coercion has replaced persuasion as means of gaining consensus.

Anti-abortion evangelicals typically follow the lead suggested by Francis Schaeffer that abortion become a litmus test for fellowship ([49]; [18]). The trustees of the Christian Life Commission narrowly defeated an effort to prohibit any pro-choice person from appearing as speaker on any CLC-sponsored program. And the trustees of The Southern Baptist Theological Seminary, Louisville, Kentucky, at their annual meeting in April 1990, approved a strongly worded resolution condemning abortion and urging all faculty and staff to support legislation forbidding abortion except to "save the physical life of the mother". The same meeting threatened dismissal of an ethics professor if he did not curtail his speaking and writing in favor of legalized abortion ([52]; [58]). The moralistic and authoritarian temperament of conservative evangelicalism is apparent in such efforts. Elective abortion is regarded as obviously immoral, which hardly allows room for debate or for disagreement ([61]; [6], Ch. 6; [7]). Legalized abortion is regarded as evidence of the moral decadence of society and a cause for the judgment of God upon America [11]. Opposing abortion becomes a crusade demanding Christian commitments and political zeal. Harsh actions toward those who are pro-choice are viewed as necessary elements of righteous indignation.

The anti-abortion claim that it has affinities to the Civil Rights movement of the 1960s has not gone without challenge. The lack of interest in social justice and racism has been noted by blacks [45]. Critics liken the anti-abortion movement to Prohibition ([46], p. 8), while others see parallels between radical efforts such as "Operation Rescue" and the Inquisition ([57], pp. 517-531). The increasing resistance by the American public to the anti-choice cause is evidence that it has failed to persuade the American people either of the personhood of the zygote or that legalized abortion is a social evil. Those who see First Amendment issues at stake in the abortion debate criticize the anti-abortion movement for attempting to base legislation on a sectarian dogma, namely, the personhood of a conceptus ([59], pp. 567-584).

EXPERIMENTATION, INFORMED CONSENT

Experimentation on human subjects was placed emphatically on the bioethical agenda by the Nazi pogroms. The Tuskegee scandal involving syphilitic prisoners and government testings of chemical warfare agents in Utah have shown that the issue is also part of the American moral landscape. Widespread concern has been generated in more ambiguous cases, such as those involving transplants and implants.

The Baby Fae case extended the quest for sources for people in need of heart transplants. Placing a baboon heart in a human infant raised the issue of informed consent in a new way. Might parents morally permit a newborn to be subjected to such experimentation, especially since the aim was not therapy but the advancement of medical knowledge? Whether the information provided the parents was adequate, as well as whether the pain endured by the infant was justified, were issues that surfaced. While the discussion and commentary among Baptists and Evangelicals was limited [54], the case made the issue unavoidable.

The artificial heart generated more interest than written response among evangelicals. Morally, the case was considerably different from the Baby Fae case in that competent, dying adults were the research subjects. Concerns focused upon whether the technology was sufficiently advanced to use on human subjects, upon issues of informed consent, and upon the costs associated with exotic techniques [55]. Baptist/evangelical emphases on individualism and a strong belief in the afterlife seem to provide grounds for approving any competent patient's right to engage in research and experimentation, as long as he or she is adequately informed. The concern for the patient as person and a strong theology of human sin raise suspicions about researchers' motives and support protection of the subject from unnecessary or unanticipated harm.

Calculating costs as a moral factor in experimentation runs counter to the emphasis on the value of the person. Such utilitarian calculus seems cold and non-personal. Making financial factors prevail over human well-being seems a *prima facie* violation of agape. For such reasons the evangelical concern will be to encourage medical science in developing new cures and procedures. Few would adopt a preference for miracle cures or faith healing over medical intervention ([43], p. 491). The miracles of Jesus give strong support for concerns about health and medical science. Staggering costs, maldistribution of resources, and threats to social stability will necessarily be included in moral reflections on experimental medicine. Just how the belief in the inestimable value of every person is to be translated into an ethical maxim for medicine is hardly obvious. The technological momentum is pushing for more sophisticated bioethical reasoning than the moralistic responses typically advanced by evangelicals.

Research involving intrauterine diagnosis of genetic anomalies either conjures up fears of eugenics ([41], p. 23) or is regarded as a hopeful breakthrough for dealing with genetically related diseases or disabilities ([53], ch. 6). The discussion will undoubtedly become both more extensive and intensive as evangelicals respond to the Human Genome Project and its implications for the future of medicine. A major revolution is in the making, involving a shift in paradigms from the Freudian emphasis on nurture as the primary factor in personality development, behavior, and health. The genetic revolution will challenge that assumption, looking more to nature and the mysteries of DNA. Unfortunately, the tendency to think moralistically about the consequences for abortion decisions where deformity is discovered tends to sideline discussions about the values of genetic research.

Fetal cell transplants are a case in which concern for the fetus as "person" conflicts with concern among anti-abortion evangelicals for the health of the diseased and afflicted [27]. These evangelicals are reluctant to give moral support for using tissue from aborted fetuses to treat people suffering from such problems as diabetes, Parkinson's disease, Alzheimer's, spinal cord injuries, and possibly schizophrenia. Research seems to indicate that these patients could benefit – perhaps even be cured – by this procedure.

A piety of "suffering as virtue" is often linked to beliefs about the fetus as person, undermining evangelical support for fetal cell therapy. Anti-abortionists see such procedures as a way to promote the "social" or "redeeming value" argument for abortion, and thus a ploy to make abortion more acceptable as a social practice. Or, they might argue that suffering is an enigma only God can resolve, preferring to leave the dying without treatment than to compromise the "life of the unborn". Some evangelicals insist that abortion is a separate issue; they can be prolife and support the use of fetal tissue transplants. Others want to require that only spontaneously aborted fetuses be used, believing that "nature" has already discarded them thus making the moral argument moot.

Anti-abortion sentiment has thus far been successful in hampering research involving fetal cell tissue. There are numerous voices calling for the moratorium to be lifted, however, believing that crucial medical advances should not be subordinated to the abortion debate ([60], pp. 24-28). Both prudence and the health interests of those who are undeniably persons require that these voices of common sense and concern prevail.

EUTHANASIA

Evangelical interest in issues pertaining to elective death has provoked a number of writings. The focus is both on the commandment forbidding killing (*Exod.* 20:13) and a theology of suffering, death, and the afterlife. The cases of Baby Doe, Karen Anne Quinlan, and William Bartling have, in many ways, symbolized the concerns with infanticide, brain death, and the right to refuse treatment. Interest in the issues has not assured depth in discussions, however, since arguments are often cast at the level of rules ([53], pp. 39-41).

Care of Severely Handicapped Newborns

Infanticide is a term often used both to reflect and evoke a moral revulsion against assisting the death even of an infant born dying. Davis, for instance, lumps all cases together, while recognizing that the life of some infants cannot be saved ("no known medical solutions") and should not be sustained by extensive intervention. The hope is for "an early and merciful death by natural causes", he says ([6], p.173). He does not struggle with the morality of leaving such infants to die a slow and ugly death. He lapses into the piety of "hope" for quick death by "natural causes". Dying naturally, however, is often a slow, painful, and degrading process. Davis deals with infanticide and abortion as simple moral issues: both are to be rejected.

The most thoughtful, thorough, and courageous treatment of this issue has been done by Earl Shelp [50]. He tackles the questions of who is to make the decision and what decision might be morally justifiable. The parents, he maintains, are the proper authority for the decision. Federal regulations (as in the Baby Doe case) are thus ill-advised, medically problematic, and morally imperious. His analysis of the morality of deciding to "allow to die" vs. that of "killing" is both careful and insightful. He questions most resolutions of the debate in terms of the option to "sustain life" as morally untenable. For the most severely impaired, "passive euthanasia" is inhumane and morally untenable. He concludes that providing a merciful death to these unfortunate children is morally licit ([50], p. 174).

Shelp's treatment struggles with the moral questions posed from the same sources of authority as does Davis. But Shelp overcomes the moralistic approach that often parades data as a camouflage for refusing to analyze the issue ethically. He is not plagued by the need to support a morality that was fashioned in a pre-technological era. A generation ago, such decisions were not subject to the interventions of imperious pressure groups or the impositions of a vast array of technological apparatus that often treats but cannot cure.

Withholding/Withdrawing Treatment

The dying of the comatose and elderly is often treated with the norm of the sanctity of life ([6], p. 174; [16], p. 154). Evangelicals are suspicious of if not hostile to active euthanasia or mercy killing. Developments in medical technology have created a crisis for moral understandings, since many things can now be done to extend the life-span or delay the death of the person. Ironically, Davis blames the "public's insatiable demand" for medical technology as the root cause of the problem, while criticizing the widespread move to execute living wills! Both Davis and Geisler ([17], p. 170) believe that there is an absolute moral distinction between "keeping alive" and "assisting to die", or between comforting and killing [68]. Suicide to avoid suffering and debilitation is regarded as "hedonic" ([40], p. 14). Never would they cross the line actually to shorten the dying process. No matter how negative the prognosis, the patient is also to be fed and hydrated [36]. The slippery slope argument, that accepting one form of assisted dying will open the door to killing other categories of people such as the senile, the comatose, and the economically burdensome, is also widely used ([6], p. 182).

Major differences appear at this point, since others believe that both euthanasic suicide and assisted dying or active euthanasia may be morally permissible. A major variable on this issue is that of the *definition of death*. Those most opposed to "euthanasia" do not deal with the issue. The morality of assisted dying is based on either the notion of whole brain death ([4], p. 125) or the absence of neocortical function ([53], pp. 131-135).

Evangelicals and Baptists are virtually unanimous in supporting the right to refuse treatment when death is imminent. Further, there is no moral obligation to sustain life artificially when there is no reasonable hope for recovery or the patient is comatose or might otherwise reasonably be considered dead ([17], p. 171; [4], p. 127; [53], p. 145).

Biblical perspectives on suffering and death are important for evangelicals in the euthanasia debate. Theologically, however, wide differences of opinion can be found. Some believe death is "unnatural" as the consequence of sin ([6], p. 182), while others hold that it is a creation of God, the experience of which has been tainted by sin ([17], p. 168; [39], p. 295; [53], p. 135). The notion of "death as evil" is paired with the belief in the sanctity of life. The moral mandate becomes a simple application – preserve life, defeat death.

A more nuanced view of the evil that may accompany the experience of death allows and requires a more critical struggle with the moral issue of whether people are simply to be passive in the face of death. It is difficult, if not impossible, to sustain either the notion of death as evil or the idea of the sanctity of life on biblical grounds. Using the latter to settle the questions posed by advancing medical technology and the maintenance of patients far beyond the point at which death would typically have come is terribly problematic.

Two critical moral questions need to be addressed. First, what are the biblical/theological grounds for the right to refuse treatment? Second, might it not be more morally responsible and loving to assist a patient's dying painlessly and quicker rather than to adopt the morality of passive waiting for death to happen "naturally"? On these questions the literature is sparse among evangelicals.

Biblical anthropology is at the heart of the debate about elective death. Anti-euthanasia evangelicals seem more concerned to constrain by law [44] what they regard as the impulse to kill than to discover the moral imperative of love in exceptional and tragic circumstances. A strong case can be made on Scriptural grounds that being made *imago Dei* requires not only anticipating death but participating in decisions about dying. Knowing that death belongs to human finitude is linked to the anticipatory and reflective capacities of being human. Knowledge about the causes and processes of death and the ability to calculate the course of the dying process underscore the moral responsibility for acting on the basis of such knowledge. Advocating passivity toward whatever happens by disease, accident, or medical technology seems more fatalism than faith. The question on the moral frontier is whether the artful uses of medical science now should be used in assisting death ([50]; [53]). Whether such "killing" is a "countervailing requirement of agape" or a malevolence disguised as mercy will generate considerable debate.

HEALTH CARE

The sensitivity to social issues among evangelicals and their interest in persons are indicated by the emergence of concern about health care in America. The 1990 "Christianity Today Institute" ([10], pp. 27-38) was devoted to the health care crisis, prompted by the American College of Physicians' call for a nationally funded health program.

G. Timothy Johnson examined the relation of health care to spirituality. An interview with C. Everett Koop, hero to many evangelicals because of his service as Surgeon General during the Reagan administration, warned against "socialized medicine". Kenneth Vaux examined "The Church as Care Giver", and F. Burton Nelson asked "What Churches Can Do".

Four proposals for health care reform at both federal and state levels were outlined, including the "Basic Health Benefits for All Americans Act (BHB)" and "The Oregon Basic Health Services Act". The latter includes a type of "allocation" of scarce resources by refusing to pay for certain types of exotic and expensive services such as heart transplants. This type of rationing was criticized by Koop as discriminatory.

Many evangelicals will be reluctant to support national health care insurance, or other types of federal funding, fearing "socialized medicine". The wedding of Christian rhetoric with laissez-faire capitalism became commonplace in the literature of the New Right, especially influential through Jerry Falwell and the Moral Majority and the "Christian Success" seminars of Tim LaHaye. Socialism and communism were favorite targets of this right-wing coalition. It will be difficult for them to overcome the highly emotional ideological bias of New Right politics to begin to think realistically about needed changes in the health care system.

The biblical and ethical principles that motivate the social sensitivities of evangelicals provide a positive framework for religious support of a national health care system. One writer contrasted his experience with the Canadian health system to that of the United States, arguing that the Christian concern should not be over whether it is public or private. "The ethical distinction", he argued, "is one of ethos and motive – between health care as ministry and health care as business" ([70], p. 22). He argued rightly that Christians center their concerns on the principle of love of neighbor, which requires mercy for every citizen. The question of expense to the national treasury he regards as irrelevant to the moral obligation to provide health care.

The issue is very much on the national agenda for the nineties, and will undoubtedly evoke heated discussions. The strong moral orientation of evangelicals to person-centered values will assure widespread support for a creative and effective health care system.

AIDS (ACQUIRED IMMUNODEFICIENCY SYNDROME)

The AIDS epidemic has sent tremors of fear and panic through the American public, tapping an enormous reservoir of sexual repression and death anxiety. Evangelicals have been slow to respond to the crisis in terms of a consistent ethic, but have been vocal in addressing certain dimensions of the issue.

Attitudes about AIDS are linked to assumptions regarding human sexuality, theological beliefs regarding God as judge, Christian conversion, and the mission of the church. On this subject as on others, evangelicals have a wide diversity of opinion. At least five responses to the victims of AIDS can be discerned: rejecting and punitive; rejecting and fearful; apathetic and indifferent; accepting, and ministering without approval; and accepting, approving, and ministering without condemnation.

A call for churches to develop a policy on AIDS that would be publicized and used as a guideline in dealing with those who are HIV positive illustrates the problem. When people panicked, a five-year-old was banned from attending Sunday school at the Moody Church in Chicago. Not having an AIDS policy in place resulted in "public relations damage", one writer says. He calls for the use of rubber gloves, disinfectants, and frequent washing of hands when working with the infected.

This writer's article is a troubling approach to the issue, since it perpetuates many of the myths upon which panic reactions depend. This far into the AIDS epidemic, such ignorance should not be so prominent. The best value of the article is its reference to Americans for a Sound AIDS Policy (ASAP), which has a federal grant to help churches address the issue [15].

By far the most vocal response has been among those most opposed to gay rights. A type of natural law ethic is adopted that makes heterosexuality normative for people and finds homosexuality a morally perverse choice defying God's created order. Biblical passages that condemn certain homosexual practices and a strong theology of suffering as the wrath of God are also stressed. A notion of disease as punishment for sin and of health as a result of virtue frames any discussion of this health care issue ([8], p. 14).

Falwell's early statements on the issue summarized and expressed this approach. AIDS is related to homosexuality, which is a "perverted lifestyle" ([13], p. 114), and the epidemic is the judgment of God on a sinful nation ([11], p. 4). A rejecting and punitive response is required, since "toleration" of homosexuality leads to God's punishment ([8], p. 14; [12]).

Southern Baptists are deeply influenced by this theology, as evidenced by the resolution passed in 1987. AIDS was seen as "a major health threat" because of a rejection of "biblical standards of decency and morality". The rule of "God's laws of chastity before marriage and faithfulness in marriage" was affirmed. Opposition was expressed to the distribution of condoms and the notion of "safe sex". The resolution ended with a call for "Christlike compassion...[for] the victims of AIDS and their families". The 1988 resolution viewed homosexuality as the cause of AIDS, linked suffering to God's punishment, and called on gays to seek forgiveness for their "abomination" [2].

The Christian Life Commission has had a mixed response to the issue. Prior to its shift to the right, it attempted to educate Southern Baptists. The Surgeon General's Report on AIDS was distributed by the CLC, and helpful treatments of the topic appeared in its journal [31]. Fundamentalists in the Convention reacted strongly to the Koop report, and information on AIDS was dropped as a regular feature of the newsletter.

Four pamphlets have since been produced, in which strongly moralistic approaches to sexuality are combined with information about AIDS and suggestions for ministry [32]. A conference was sponsored by the CLC in February 1990, hearing from people with AIDS, family members affected by AIDS, pastors, and other professionals who minister to persons struggling with the crisis.

An ethic of compassion based upon Jesus' response to lepers and the commandment to love the neighbor constantly temper the responses of most Baptists [51]. Open acceptance of homosexuality is difficult to find, but there is both openness to and acceptance of those suffering from AIDS. One local church was ousted from a California association when it called verbal gay-bashing over the AIDS crisis "disgusting" and "unchristian" ([19], p. 3). An autobiographical account of a man who died of AIDS is widely distributed [3]. He strongly denounces (his former) homosexual activity, rejects the notion of God's punishment, and advocates an end to the categories of "innocent" or "guilty" in dealing with victims of the disease. Another book was designed to facilitate discussions in church study groups [64]. It strongly advocates ministry responses, but does so while rejecting homosexuality as a lifestyle. A Florida pastor led his congregation in a ministry of compassionate support for persons with AIDS [1], and a Southern Baptist minister is chair of the Dallas AIDS Interfaith Network [63].

The American Baptist statement recognized the health threat and the stigma attached to victims, and called for repentance for being reluctant to minister to people because of the stigma [37], p. 18).

ANIMAL RIGHTS

There is a paucity of material on the subject of animal rights, but the topic will undoubtedly merit increased attention as the issue comes to the forefront of public concern ([61]; [69]).

Tim Stafford says that the animal rights movement has a non-Christian point of view. He is critical of the notion of "speciesism", but identifies a point of contact with the advocates of animal rights in the fact that human beings are special ([61], p. 20). He argues that the issue is fundamentally religious, and that the great mistake of liberationists is the attempt to exclude religion. He also charges that animal rights activists do not present a clear or convincing rationale; they argue from emotion or feeling rather than reason (p. 23). He concludes by admitting a wide spectrum of thought among Christians from the (Hindu) sanctity of all life to the notion that all non-human creatures are given to people for their use. He does suggest a positive note from *Psalm* 104 that portrays animals as part of God's creative and joyful kingdom. Care for animals is part of Christian calling, he says. His hope is that post-Christian society will recover a biblical framework for thinking about ecology, suffering, life and death both for people and animals.

Southern Baptist Theological Seminary
Louisville, Kentucky, U.S.A.

BIBLIOGRAPHY

1. Amos, W.E., Jr.: 1988, *When AIDS Comes to Church*, Westminster, Philadelphia.
2. *Annual of the Southern Baptist Convention*.
3. Arterburn, J: 1988, *How Will I Tell My Mother?*, Oliver Nelson, Nashville.
4. Barnette, H.H.: 1982, *Exploring Medical Ethics*, Mercer University Press, Macon, Georgia.
5. Childress, J.F., and J. Macquarrie (eds.): 1986, *The Westminster Dictionary of Christian Ethics*, Westminster, Philadelphia.
6. Davis, J.J.: 1985, *Evangelical Ethics: Issues Facing the Church Today*, Presbyterian and Reformed Publishing Co., Phillipsburg, New Jersey.
7. Davis, J.J.: 1984, *Abortion and the Christian*, Presbyterian and Reformed Publishing Co., Phillipsburg, New Jersey.
8. Dobson, E.: 1985, "AIDS: Looking for the Cure", *Fundamentalist Journal*, October, 4.
9. Eighmy, J.L.: 1972, *Churches in Cultural Captivity*, S.S. Hill (ed.) University of Tennessee, Knoxville.
10. "Emergency": 1990, *Christianity Today*, 34:12 (September 19), 27-38 .
11. Falwell, J.: 1980, *Listen, America!*, Bantam, New York.
12. Falwell, J.: 1987, "AIDS: the Judgment of God", *Liberty Report*, April, Lynchburg, Virginia.
13. Falwell, J.: 1987, "An Agenda for the 1980's", in J. Neuhaus and M. Cromartie (eds.), *Piety and Politics*, Ethics and Public Policy Center, Washington.
14. Fowler, P.B.: 1987, *Abortion: Toward an Evangelical Consensus*, Multnomah, Portland.
15. Frame, R.: 1990, "AIDS: Coming to a Church Near You?", *Christianity Today*, June 16, pp.or 50-52.
16. Frame, R.: 1990, "Time to Face the Consequences", *Christianity Today*,

pp. 48-52.
17. Geisler, N.L.: 1989, *Christian Ethics: Options and Issues*, Baker, Grand Rapids.
18. Geisler, N.L.: 1987, "Sanctity of Human Life", in K.L. Kantzer (ed.), *Applying the Scriptures*, Zondervan, Grand Rapids.
19. Godges, J.: 1986 "Healing a City: AIDS in San Francisco", *Second Opinion*, p. 3.
20. Henry, C.F.H. (ed.): 1973, *Baker's Dictionary of Christian Ethics*, Grand Rapids, Baker.
21. Henry, C.F.H.: 1989, "What Does It Mean to be Evangelical?", *Christianity Today*, June 16.
22. Henry, C.F.H.: 1986, "Evangelical Ethics", in J.F. Childress and J. Macquarrie (eds.), *The Westminster Dictionary of Christian Ethics*, Westminster, Philadelphia, p. 212.
23. Henry, C.F.H.: 1986, "Fundamentalist Ethics", in J.F. Childress and J. Macquarrie (eds.), *The Westminster Dictionary of Christian Ethics*, Westminster, Philadelphia, p. 241.
24. Hinson, E.G.: 1981, "Baptists and Evangelicals: What is the Difference?", *Baptist History and Heritage* XVI (April), Historical Commission, SBC, Nashville.
25. Hoffmeier, J.K. (ed.): 1987, *Abortion: A Christian Understanding and Response*, Baker, Grand Rapids.
26. Kantzer, K. (ed.): 1987, *Applying the Scriptures*, Zondervan, Grand Rapids.
27. Lawton, K.A.: 1988, "Fetal Tissue Transplants", *Christianity Today*, March 18.
28. Lawton, K.A.: 1988, "Hassle-Free Abortion", *Christianity Today*, December 9.
29. Lawton, K.A.: 1990, "Scanning the Abortion Battlefields." *Christianity Today*, June 16.
30. Lewis, Larry L.: 1987, "'Yes' to the Human Life Amendment", *Word and Way*, January 15, Jefferson City, Missouri.
31. *Light*, 1988, March/April.
32. *Light*, 1990, April/June.
33. *Life in the Balance*: 1989, Christian Life Commission (SBC) Seminar Proceedings, Nashville.
34. McGee, D.B.: 1988, "Issues of Life and Death", in W.K. Tillman, Jr. (ed.), *Understanding Christian Ethics*, Broadman, Nashville.
35. McQuilkin, R.: 1989, *Biblical Ethics*, Tyndale House, Wheaton, Illinois.
36. Maudlin, M.G : 1989, "Last Rights", *Christianity Today*.
37. Melton, J.G. (ed.): 1989, *The Churches Speak on AIDS*, Gale Research, Detroit.
38. Mills, W.E., *et al.* (eds.): 1990, *Mercer Dictionary of the Bible*, Mercer University Press, Macon, Georgia.
39. Moody, D.: 1981, *The Word of Truth*, Eerdmans, Grand Rapids.

40. Neff, D.: 1990, "The Suicide Machine", *Christianity Today*, August 20, p. 14.
41. Neff, D.: 1990, "The Eugenic Temptation", in *Christianity Today*, November 19.
43. Numbers, R.L., and D.W. Amundsen (eds.): 1986, *Caring and Curing: Health and Medicine in the Western Religious Traditions*, Macmillan, New York.
44. Pagels, E.: 1988, *Adam, Eve and the Serpent*, Random House, New York.
45. Perkins, S.: 1989, "The Prolife Credibility Gap", *Christianity Today*, April 21.
46. Pierard, R.V.: 1989, "Wanted: Consensus on Abortion", *Christianity Today*, October 6, p. 8.
47. Pyle, H.: 1990, "Sword of the Lord Rejects New Evangelicalism", *Sword of the Lord*, September 19.
48. Pyle, H.: 1990, "Southern Baptists are Not Fundamentalists", *Sword of the Lord*, July 20.
49. Schaeffer, F., and C.E. Koop: 1979, *Whatever Became of the Human Race?*, Fleming H. Revell Co., Old Tappan.
50. Shelp, E.E.: 1986, *Born to Die? Deciding the Fate of Critically Ill Newborns*, Free Press, New York.
51. Shelp, E.E., and R. Sunderland: 1987, *AIDS and the Church*, Westminster, Philadelphia.
52. Sider, R.J.: 1987, *Completely Pro-Life*. Intervarsity Press, Downer's Grove, Illinois.
53. Simmons, P.D.: 1983, *Birth and Death: Bioethical Decision Making*, Westminster, Philadelphia.
54. Simmons, P.D.: 1985, "Baby Fae: Moral Concerns in Christian Perspective", *Western Recorder* (February 26), Middletown, Kentucky, and *Word and Way* (January 24), Jefferson City, Missouri.
55. Simmons, P.D.: 1986, "Ethical Considerations of Artificial Heart Implantations", *Annals of Clinical and Laboratory Science* 16:1.
56. Simmons, P.D.: 1987, "'No' to the Human Life Amendment", *Word and Way*, January 15, Jefferson City, Missouri.
57. Simmons, P.D.: 1989, "The Pastor as Prophet: How Naked the Public Square?", *Review and Expositor*, pp. 517-531.
58. Simmons, P.D.: 1990, *Personhood, the Bible and the Abortion Debate*, rev. ed., RCAR, Washington.
59. Simmons, P.D.: 1990, "Religious Liberty and the Abortion Debate", *Journal of Church and State*, pp. 567-584.
60. Simons, A.: 1990, "Brave New Harvest", *Christianity Today*, November 19.
61. Stafford, T.: 1989, "The Abortion Wars", *Christianity Today*, December 16.
62. Stafford, T.: 1990, "Animal Lib", *Christianity Today*, June 18, p. 21.

63. Tapia, A.: 1987, "High Risk Ministry", *Christianity Today* August 7.
64. Tillman, W.: 1990, *AIDS: A Christian Response*, Convention, Nashville.
65. Trull, J.: 1987, "Where Do Evangelicals Stand on the Issue of Abortion?", *Choosing Life: Southern Baptists and Abortion*. Proceedings of the 1987 National Christian Life Commission (SBC) Conference, Nashville.
66. Waddle, R.: 1990, "Baptist Agency Vows Active Abortion Fight", *The Tennessean*, September 13, Nashville.
67. Wallis, J.: 1980, "What Does it Mean to be Pro-Life?", *Sojourners*, November.
68. Wennberg, R.N.: 1989, *Terminal Choices: Euthanasia, Suicide, and the Right to Die*, Eerdmans, Grand Rapids.
69. Wilkinson, Loren: 1990, "A Theology of the Beasts", *Christianity Today*, June 18, p. 21.
70. Wilson, Everett L.: 1990, "When Mercy Becomes a Business", *Christianity Today*, February 19, p. 22.

AVRAHAM STEINBERG

JEWISH MEDICAL ETHICS

I. INTRODUCTION: THE FOUNDATIONS OF JEWISH LAW

Halakhah is the generic term for the whole legal system of Judaism, embracing all the laws, practices and observances of Judaism. Orthodox Jews consider the *Halakhah*, in its traditional form, to be absolutely binding. The source of *Halakhah* is divine revelation. To the basic corpus of Biblical law were added rabbinic enactments and decrees. The sources of authority in *Halakhah* are composed of different elements: a) the written law, a composition of 613 positive and negative commandments of Sinaitic origin, which are included in the books of the Pentateuch; b) the oral law, which includes the interpretation of the written law transmitted in its entirety with its details and minutiae at Sinai, as well as logical deductions, and also rabbinical decrees, customs, positive enactments, and negative enactments. A distinction exists between laws of Sinaitic origin, termed *de-orayta*, and laws of rabbinical source, termed *de-rabbanan*.

The development of *Halakhah* may be divided into six periods.

- The earliest and most important source is the Bible.
- The Tannaitic period (1-220 C.E.): The major work of this period is the *Mishnah*, edited by R. Judah ha-Nasi at the end of the second century, in which were summarized all the legal debates and decisions of the scholars up to that era. This work became a sacred text second only to the Bible.
- The Amoraic period (220-470 C.E.): The *amoraim* were the interpreters of the *Mishnah*, and their work is termed *Gemarah*, in which discussions and comments on the *Mishnah* constitute the major part; but also included are numerous new issues of legal importance, as well as folklore, history, astronomy, geography, botany, zoology, and medicine. The combination of *Mishnah* and *Gemarah* is termed *Talmud*. The Palestinian *amoraim* collected their work in the *Jerusalem Talmud*, and the Babylonian *amoraim* in the *Babylonian Talmud*. With the closing of the *Talmud*, this work virtually became the infallible source of the *Halakhah*, and the history of post-Talmudic *Halakhah* is founded on recourse to the *Talmud* as the final and overriding authority. Of the two *Talmuds* the Babylonian became the more authoritative. Rules for determining the actual decision in law from the labyrinth of legal debate in the *Talmud* have been formulated.
- The Geonic period (7th century to mid-11th century): This period is marked by a significant expansion of the *Talmud* among Jews of the world, further analysis and deeper comprehension of the *Talmud*, and the beginning of

formal codification.

- The period of the *Rishonim* (mid-11th century to 16th century): The *Rishonim* were divided into commentators, whose major activity consisted of exposition of the classical texts of the *Talmud* and other early rabbinical works, and decisors, whose major activity was to formulate and finalize legal conduct. The works of the decisors are divided into *responsa* and codification. The greatest Jewish codifier was Maimonides (1138-1205), who compiled his gigantic code, the *Mishneh Torah* (called after his death the *Yad ha-Chazakah*), in which he presented the final decisions in all matters of *Halakhah*, including those laws which were no longer applicable in his days. The other major work of codification of the Jewish law was done by Joseph Karo (1488-1575) in his *Shulchan Aruch*, which marked a turning point in the history of *Halakhah*. Adherence to the *Shulchan Aruch* became the test of Jewish Halakhic fidelity. *Responsa* is a rabbinic term denoting an exchange of letters of queries and replies in which one party consults another on a Halakhic matter. The beginning of *responsa* literature as a literary and historical form of important dimensions took place in the middle of the Geonic period, and has continued to this day. This form of Halakhic discourse has served as the major opus of normative Jewish legal formulation. It is estimated that there are about 300,000 *responsa* collected in over 3,000 books of *responsa*.
- The period of the *Acharonin* (16th century to present time): A similar distinction has continued between commentators, codifiers and responders [14].

II. THE PRINCIPLES OF JEWISH MEDICAL ETHICS

Over the past three decades a significant number of books (heralded by Jakobovits [33]) and periodicals (initiated by Steinberg [45]) have been published that are entirely devoted to Jewish medical ethics (see Appendix). In recent years, Jewish medical ethics has yielded an ever-growing body of discussion, writings, and research on relevant Jewish values and laws, so that a voluminous body of literature now exists in this field.

Jewish medical ethics, in the sense of normative Jewish law and philosophy applied to practical medical issues, is characterized by and distinguished from secular medical ethics in its range, in its applications, in its methods of analysis, and in its endpoint.

With respect to its range, Jewish medical ethics deals with all the diverse universal current ethical problems in medicine, but also includes "pure" Halakhic topics related to the practice of medicine, which are no less important from the Jewish standpoint. Examples of "pure" Halakhic issues include the proper balance between medical intervention and desecration of the Sabbath and holidays; medical aspects of circumcision; the specific problems of a priest (*Cohen*) as a patient, physician, visitor of the sick, or hospital employee, and the prohibition against his contact with the dead; Jewish dietary laws (*Kashruth*) of medications and food for the sick; and medical interventions in menstruating women (*Nidah*).

With respect to underlying principles, many of the Jewish moral and legal principles and rules have significant bearing upon the practice of medicine. Obviously, they cannot be described in any detail in this report. It is, however, worthwhile to mention a few fundamental principles, since a deep rift exists between current Western secular and Jewish moral foundations concerning medicoethical issues:

- Judaism subscribes to commitments, obligations, duties and commandments rather than to rights and pure hedonism; beneficence and altruism are promoted over mere nonmaleficence.
- Judaism is opposed to absolutizing any single precept; rather, a golden path, a middle way, is always advocated and encouraged. When conflicting values in practical medicine are encountered, each patient must be considered individually, a casuistic approach is applied, and a solution is reached depending on the specific clinical and ethical circumstances. This is, obviously, done within the general framework of Halakhic rules and regulations.
- The principal aim of studying ethics and Jewish law is to act accordingly, not merely to engage in intellectual exercise or academic analysis.
- Judaism subscribes to a higher order of moral conduct than secularism as obligating the individual and society. Autonomy as a concept of respect for others is highly valued and demanded. However, autonomous decisions that do not comply with the required moral standard are overridden by higher moral values, as determined by *Halakhah,* which govern the life of each individual, patient and physician alike. Judaism restricts the notion of autonomy to actions that are Halakhically indifferent. Where conflicting values arise, each individual is bound to act in accordance with the required normative moral conduct. Thus, everyone is duty-bound to act according to that standard and to relinquish his or her temporary wishes. This approach could be termed a moral-religious paternalism, in contrast to the Hippocratic individual-personal paternalism of the physician.

The physician-patient relationship is viewed by Judaism as a covenant, which does not provide for the notion of freely contracting individuals. There is an obligation upon a physician always to extend his help to those in need of his services: "The Torah gave permission to the physician to heal; moreover, this is a religious precept and is included in the category of saving life; and if the physician withholds his services it is considered as shedding blood" (Karo's *Shulchan Aruch*, *Yore De'ah* 336:1). According to Judaism, the physician is God's messenger in healing people in need. The physician's obligation to save life is independent of the patient's wishes or opposition.

Judaism views the seeking of medical attention as a moral imperative. Moreover, man is obligated to exercise prudence in preserving health. No one has the right to refuse medical treatment deemed necessary and effective by competent opinion. In the Pentateuch it is stated: "Duly take heed to yourself, and keep your soul diligently" (*Deuteronomy* 4:9); "Take you, therefore, good heed unto yourselves" (*Deuteronomy* 4:15). From these and other scriptural phrases, the Talmudic sages (*Babylonian Talmud, Tractate Berachot* 32b) and the decisors deduced the obligation to protect our bodies from harm and danger. Maimonides (*Misneh Torah, Rotzeach U'Shmirat Ha'NeFesh* 11:4-5) and Karo (*Shulchan Aruch*, end of *Choshen Mishpat*) write: "The sages prohibited many things because they involve danger to life. Whoever disregards these things and their like and says: 'I will place myself in danger; what concern is it to others?', or 'I am not particular about such things' – disciplinary flogging is inflicted upon him. But he who is careful about these matters will receive great blessing". According to some rabbinic decisors and philosophers, the source for the obligation of preserving one's health and life is a general principle and a fundamental obligation of preserving the entire *Torah*. For instance, "The reason why the *Torah* exhorts one about the preservation of health is because the Holy One, blessed be He, created the world with His benevolence to benefit His creations so that they recognize His greatness and serve Him by fulfilling His commandments...[H]e who endangers himself is as if he shows contempt for the will of his Creator and does not want to serve Him or reap His reward. There is no greater disrespect than such despisement" (Rabbi Moses Rivkas (17th century), *Beer Ha'Golah* commentary at the end of Karo's *Shulchan Aruch*, *Choshen Mishpat*). Or again, "It is a commandment to conduct oneself with good characteristics and good habits to preserve one's health in order to be healthy and strong to serve the Creator, may He be exalted" (Rabbi Israel Meir Ha'Cohen (1839-1933), *Mishnah Berurah* commentary on Karo's *Shulchan Aruch*, *Orach Chaim* 155:11) (for detailed references see [44], pp. 95-98).

With respect to methods of analysis, the foundation of Jewish medical ethics is the Sinaic revelation and the interpretation of the law by authorized and recognized scholars (rabbis or decisors). The mode of discourse and the cognitive background pertaining to Jewish medical ethics is based on the ancient sources of the Bible, the *Talmud*, and the vast literature of codification and *responsa*. Any ethical issue in medicine is analyzed according to the foundations of *Halakhah*, much in the same way as any other issue in life when a Halakhic problem arises.

With respect to its endpoint, Judaism, like other religions and the secular law, gives finite and decisive answers to the ethical issues posed to the decisors, although the solutions are not always accepted by all members of the community. Dispute-resolution mechanisms are built into the Halakhic system to finalize the actual conduct. The Jewish model of decision making in medicine is composed of a triad: the physician, who has the obligation to treat the patient and to recommend to him the best medical advice; the qualified rabbinic authority, who is required to solve any ethical or legal problem that may be encountered by either the patient or the physician; and the patient, who has the obligation to seek medical help, and has the autonomy to choose his medical and Halakhic advisers, and to decide on matters that do not involve either medical or ethical expertise (for general reference see [47]).

III. PARTICULAR BIOETHICAL ISSUES

A. IN VITRO FERTILIZATION (IVF)

Judaism does not recognize a doctrine of natural law as such. Judaism teaches that nature was created by God for man to use to his advantage and benefit, and doing so is viewed as a positive partnership between man and God to improve nature for the benefit of man. Hence, in the absence of a specific biblical or rabbinical prohibition, man is free to utilize scientific knowledge in order to overcome impediments of nature. Therefore, performing IVF is not considered an interference with God's will; rather, it is within the realm of improving the world and helping a fellow human being to overcome the impediment of infertility.

Moreover, the new reproductive techniques do not pose any contradiction to the fundamental beliefs in Judaism, since they do not create a basic change in nature; rather they represent an outstanding technical improvement. An important aspect is that they do not change or solve the mystery of life, which Judaism believes is made possible only by the Creator. Hence, performing IVF is not considered as tampering with life itself.

Although most contemporary rabbis rule that there is no Jewish legal objection to IVF using husband's sperm, no woman is required to submit to this procedure, since the obligation to populate the world is limited to bearing children by means of natural intercourse ([10], p. 88). In any case, women in general are exempted from the obligation to be fruitful and multiply. However, if medico-psychological reasons stand behind the request of a couple to use the technique of IVF, it should be considered a legitimate medical intervention [48].

The means employed in procuring sperm for purposes of IVF pose a Halakhic problem. Jewish law forbids ejaculation other than within the context of intercourse as destruction of the seed for naught. The question is whether or not semen procurement designed to promote procreation is deemed to be for naught. Most Halakhic authorities sanction the practice of sperm and egg procurement for the procedure of IVF, because the aim is to fulfill the Biblical commandment of procreation.

Most contemporary rabbis approve IVF if a husband-wife team contributes sperm and egg with the following qualifications: only if the sperm of the husband and the egg of the wife are used exclusively, proper safeguards are established to assure that no sperm other than the husband's will be used, and no other way is available to solve the couple's infertility problem ([10], Ch. 15-

16); [36]; [77]). (This procedure was also endorsed by the Chief Rabbi of Jerusalem, permitting The Shaare Zedek Medical Center in Jerusalem to perform IVF under these conditions.) However, a few dissenting opinions prohibit IVF even between husband and wife ([49]; [67]).

Legal issues from the Halakhic standpoint have been raised: Does a filial relationship exist between the father and a child born in this manner? Is the child considered to be an heir to his father's estate? In any event, the resolution of these questions has no bearing upon the permissibility of IVF ([10], Chapters 15-16).

With respect to IVF which uses gametes contributed by a donor outside the marriage, a distinction exists between the use of donor sperm and donor egg. If one uses sperm other than that of the husband, there are many objections to this procedure. This act is considered by most rabbinic opinion to be an "abomination" and strictly prohibited for a variety of reasons, including the possibility of incest (the child born of such IVF may later marry a sibling unknowingly), lack of genealogy (the father's identity is unknown), and problems of inheritance (does the child inherit the genetic father, the adopted father, or both?). A few rabbis regard such insemination as adultery, requiring the husband to divorce his wife. Most rabbis, however, state that without a sexual act involved, the woman is not guilty of adultery. Regarding the status of the child, rabbinic opinion is divided. Most consider the offspring to be legitimate, but a minority of rabbis consider the child to be an illicit child. Hence, under no circumstances should the sperm of any person other than the husband be utilized ([38], Chapter 9).

Surrogate motherhood via a donor egg is characterized by contemporary Halakhic scholars as offensive to moral sensitivities when resorted to as a convenience in order to avoid the encumbrances of pregnancy. To use another person as an "incubator" and then to take from her the child she has carried and delivered for a fee is a revolting degradation of maternity and an affront to human dignity ([33], pp. 244-250, 261-266). However, if a donor egg or a donor womb was utilized, a fundamental issue arises: Who is the legal mother of the born child? How is motherhood defined – by nature, genetically, so that the woman who gives the fetus its genetic make-up and characteristics is the legal mother, or by nurture, so that the woman who contributes to the acquired characteristics of the child and creates mother-fetus bonding is considered the legal mother? Determining the maternity of a child born of a fertilized ovum implanted in the womb of a host mother is still being debated in contemporary sources. In Jewish law, the definitions concerning genealogy are not necessarily based on laws of genetics. Hence, based on Talmudic and midrashic analogies, most rabbis state that, according to Jewish law, the maternity of the child is determined by the mother who nurtures and gives birth to the child, rather than by the biological mother. A minority of rabbis conclude that the biological mother is the legal mother. Some rabbis consider both women to be mothers of the child [48]. Nonetheless, if a married woman becomes a host mother via a donor egg, the husband would certainly not have to divorce his wife, since no act of adultery was committed.

If one obtains several eggs from the mother's ovary and fertilizes all of them, is one permitted to destroy those that will not be reimplanted? Is one permitted to perform medical experiments on unused fertilized eggs? Some Jewish scholars have expressed the opinion that there is no human fetal life outside the uterus; hence a superfluous fertilized ovum can be disposed of ([28]; [38], Chapter 9).

B. ABORTION

Abortion as a medical procedure is not mentioned in the Bible. In the *Talmud* and in the literature of the early decisors this issue is only sparsely discussed and defined. The reason for this, most likely, is that abortion was not performed by Jews in those days ([72], p. 23). The issue of abortion has, however, received great attention in the last two centuries, and a voluminous literature concerning all aspects of abortion is available in the *responsa*.

Three possible approaches exist toward abortion: (a) virtually absolute pro-life, whereby abortion is prohibited without any exception; (b) virtually absolute pro-choice, whereby the pregnant woman can choose to abort her fetus at her will, without any explanation; (c) a golden, in-between pathway, whereby abortion is prohibited in principle, but may be sanctioned when appropriate reasons exist. The third is the principle by which Judaism solves the issue of abortion. All Jewish authorities agree that abortion is forbidden for various reasons. All agree that feticide is not punishable as a capital crime. All agree that the life of the mother takes precedence over that of the unborn child (*Mishnah Ohalot* 7:6), because the fetus's title to life is subordinate to that of the mother, or because of a "pursuit" argument, whereby the fetus is "pursuing" the mother [48]. Hence, it is accepted by all Jewish scholars that abortion is prohibited when performed merely as an act of the woman's autonomous decision, or when considered for minor inconveniences, or for socio-economic or educational reasons. It is also accepted by all Jewish scholars that abortion is sanctioned when the life of the pregnant woman is endangered by the continuation of her pregnancy. According to most authorities, the performance of an abortion may be warranted to preserve the mother's physical health (e.g., because of cardiac or renal disease) or mental health (e.g., because of psychiatric depression).

However, several issues are still debatable. Some authorities maintain that the performance of abortion is rabbinically prohibited, whereas most Jewish scholars maintain the prohibition to be of biblical origin (see [48]; [55]; [58]; [74]. The exact source of the biblical prohibition is also debatable. Some reason that interference with pregnancy would constitute expulsion of semen for naught, an act that is strictly forbidden in Jewish law, while other authorities maintain that the ban against abortion is because of illicit wounding of the woman. However, the strictest views are those which consider abortion as murder, though not punishable, because the fetus is seen as a potential or a partial person [48]. The moment of ensoulment, although discussed in the *Talmud* (*Babylonian Talmud*, *Tractate Sanhedrin* 91b), has no bearing on the practical, jurisprudential issues concerning abortion ([24], pp. 271-275).

Based on the differences of opinion about whether abortion is forbidden by biblical law or by rabbinical law, there is an ongoing debate about whether it is permissible to abort a defective fetus. Those authorities who maintain that abortion is forbidden by biblical law rule that a defective fetus has the same claim for life as a normal fetus; hence abortion is not permitted for fetal indications ([7]; [16]; [53], pp. 344-376). On the other hand, those authorities who maintain that abortion is forbidden by rabbinical law accept the prevention of suffering, both on the part of the born deformed baby and on the part of the family, as valid reasons to permit abortion ([25]; [30]; [64]; [65]).

There are, however, two exceptional fetal indications for which all authorities agree that abortion is permitted. One is an anencephalic fetus, which is not considered a human being; therefore abortion is permitted [39]. The other is selective reduction of a pregnancy with multiple fetuses, based on the argument of "pursuit", whereby the fetuses are "pursuing" the mother or the other fetuses ([8]; [27]; [38], Chapter 12; [40]).

The question of the inviolate nature of embryonic life during the first forty days following conception, prior to organogenesis, is a matter of some dispute. Until forty days after conception, the *Talmud* considers a fertilized egg nothing more than "mere fluid" (*Babylonian Talmud*, *Tractate Yevamot* 69b). Some authorities maintain that fetal development during this period is insufficient to warrant independent standing in the eyes of Jewish law, whereas others stress that the fetus's potential for development into a viable human being is sufficient to guarantee it an inviolable claim for life [48].

Most authorities rule that termination of a pregnancy resulting from rape or incestuous relations is not permitted [48].

C. FETAL TREATMENT

There is an increasing legal and ethical recognition of the rights of viable fetuses. Ethical issues arise when the mother's constitutional rights of autonomy, liberty, and privacy conflict with the fetus's right to be born healthy, or to be healed from a life-threatening condition by medical or surgical intervention in the mother.

From the Jewish viewpoint, the relationship between the mother and the fetus is based on the description in the section on abortion. If the risk to the mother is very small, she is obliged to allow her fetus to be treated for a correctable medical or surgical condition. If the risk is substantial, the mother is not obligated to endanger her life to save that of her unborn fetus, because her life takes precedence over that of the fetus ([38], Chapter 13).

D. NEWBORNS WITH ANOMALIES

According to Judaism, infanticide is no less a crime than the murder of an adult person, and a handicapped newborn has the same rights as an older handicapped person. The determining factor in the decision whether or not to prolong the life of an anomalous newborn depends on his chances to survive. Of no significance, as far as treatment is concerned, is the expected quality of life [42]. Hence, a baby with a lethal lesion should be given treatments necessary to keep him comfortable, to nourish him, and to provide him with oxygen, as well as measures to reduce suffering, but he need not be resuscitated, and no heroic measures to prolong his life are required. Thus, there is no obligation to resuscitate an extremely prematurely born baby, since he has no reasonable chances to survive [71]. However, it would be right to attempt resuscitation and treatment, if a small minority is expected to survive [41]. Nonetheless, active killing of any anomalous newborn, or deprivation of basic sustenance, is absolutely forbidden, even in cases such as anencephaly ([2], pp. 207-212). This approach is identical to the attitude taken towards a terminally ill adult patient.

A newborn who has a chance to live, even if the expected quality of life is poor, should be treated in the same way that one would treat a healthy newborn. Hence, there is an obligation to treat a baby with meningomyelocele by surgically covering the defect, and correcting other dysfunctions such as hydrocephalus, urinary, and orthopedic problems. It is also obligatory to operate on a baby with Down's syndrome and duodenal atresia to save his life [63].

If the parents wish to continue all available treatment, even in babies with lethal lesions, their wish should be respected. However, if they wish to discontinue life-sustaining measures in an anomalous newborn, this should be evaluated according to the above rules, and, if needed, a court-appointed surrogate should take over the care for the baby [42].

E. INFORMED CONSENT TO TREATMENT AND EXPERIMENTATION

According to Jewish philosophy, a human being's life is not his to dispose of at will, since man is but the steward of the life that God, the Creator of Life, has given him. The Almighty entrusted the human body to the human being in order for him to keep it in good shape, to preserve it, and to nourish it. Therefore, the patient is positively obliged to seek medical help, and to do all that is necessary to preserve his life and health.

The autonomy of man is restricted to conduct that is in accordance with acts of beneficence and nonmaleficence, both to others and to himself. Hence, there is no validity to wishes that bring about destruction and harm to any one, including oneself, and the obligation to preserve life and health, being a definite halakhic duty, overrides the notion of autonomy.

Therefore, in clearcut cases where medical opinion is undivided as to the advisability of a medical procedure, treatment should be administered independently of the patient's wishes, and he may even be forced to receive the recommended treatment to save his life (Rabbi Jacob Emden (1697-1776), *Mor U'Ketziah Commentary on Karo's "Shulchan Aruch, Orach Chaim"*, 328). This is true even when a patient refuses treatment for religious reasons, such as his wish to avoid the desecration of the Sabbath, or his wish to accept suffering for the purpose of repentance.

This approach, however, is qualified by several conditions: a) the treatment is of proven efficacy, as judged by the majority of the health care providers; b) a reasonable effort has been made to convince the patient to consent to the proposed treatment; and c) no significant complications or side effects are expected. In order to avoid misuse and abuse of the authority entrusted to the physician, the coercive treatment should be limited to outstanding circumstances and be supervised by rabbis or other representatives of the public ([9]; [34]; [78]).

One is not allowed to injure, damage, or hurt oneself, since man is not the absolute master of himself. Hence, consent to injure oneself is invalid, and any decision by a person whereby he may be injured or die for reasons halakhically insufficient is waived, unless intended for an acceptable purpose. Thus, a "Jewish review board" may waive a subject's consent to undergo hazardous experiments, even when given autonomously, in instances where risk to life is untenable.

Jewish law accepts the principle that the possibility to prolong life takes precedence over the probability of hastening death if unsuccessful. Therefore, if there is some chance for cure or longer survival, even when failure of the procedure may shorten life, the patient is at liberty to undergo a therapeutic experimental modality if he so wishes; he may also legitimately refuse this treatment. This rule, however, is qualified by the following conditions: a) life expectancy without the proposed experimental therapy is shorter than 12 months [21]; and b) a majority of proficient medical experts recommend the procedure, based on their assumption that without treatment death will certainly occur [1].

It is permissible to submit to a nontherapeutic medical experiment only if it involves no risk greater than daily risks. If, however, it is not certain whether the subject's life will be endangered, and if a dangerously ill patient on hand may benefit from the experiment, there is a debate amongst Jewish scholars as to whether or not it is permissible to participate. Some halakhic authorities forbid it ([33], pp. 291-294; [60]), while others regard it as permissible ([11], pp. 384-385; [15]). The usual experimental nontherapeutic situation, however, is intended to gain general knowledge which may be helpful only in the future; in such an instance, if a significant risk is anticipated, one is not permitted to consent to it ([1]; [35], pp. 288-294).

F. CONFIDENTIALITY

Judaism places stringent restrictions upon disclosure of any confidential information. This is derived from the Biblical verses "Thou shalt not go up and down as a tale-bearer among thy people" (Leviticus 19:16) and "reveal not the secret of another" (*Proverbs* 25:9). Maimonides equates the tale-bearer to a murderer, and one who may destroy the world (*Mishneh Torah*, *De'ot* 7:1-2).

However, tension may arise between respect for privacy and the prevention of harm to an innocent person. The overriding obligation to protect others takes precedence over the value of privacy. *Halakhah* acknowledges that, in certain circumstances, even professional confidences must be revealed. Medical information may, and in fact occasionally must, be disclosed in order to prevent potential harm or damage which may occur to another innocent party ([10], Chapter 5; [13]; [37], pp. 332-335; [68]; [76]). For example, a physician is required to report an epileptic patient to authorities responsible for issuing driver's licenses [76].

This general rule is, however, restricted to the following regulations: knowledge must be firsthand, not speculation and hearsay; the information should be divulged only to those in danger or those who may act to prevent the damage; the sole motivation prompting disclosure must be the benefit to the third party, and never be based on hatred; the information should be relevant to important, not minor, issues; when disclosure is necessary, it must be limited to those facts and details which require disclosure in order to prevent harm; it should be disclosed as it is, without any exaggeration; and it should be disclosed at a time or situation at which it will be efficacious (R. Israel Meir Ha'Cohen (1839-1933), *Chafets Chaim*, *Hilkhot Issurei Rekhilut*, *Klal* 9).

Disclosure of medical information to the courts of law is mandatory and takes precedence over the notion of medical confidentiality (*Leviticus* 5:1; Karo's *Shulchan Aruch*, *Choshen Mishpat* 28:1). Although a physician, according to the Hippocratic oath, should not disclose a patient's secrets, the oath is not binding, and court may force an expert witness to provide information in order to carry out the important task of justice ([33], pp. 210-211; [37]; [61]).

G. ALLOCATION OF SCARCE RESOURCES

Judaism views society as a separate entity which has its legitimate interests beyond the sum of the interests of individuals. Societal ethics are indeed different from individual ethics. Society or government is needed to prevent man from destroying his fellow man. Therefore, society is not bound by the same ethical principles that bind individuals, such as the infinite worth of man [51].

This is based both on Jewish philosophical foundations, which ascribe great importance to society, and on the halakhic ruling concerning the redemption of captives. The *Talmud* asserts that captives should not be redeemed for more than their value in order to prevent abuse. Two reasons are given why it is forbidden to pay an exorbitant amount of money to ransom anyone: to discourage kidnappers, and to prevent society from becoming impoverished by having to divert its resources to the ransoming of captives (*Tractate Gittin* 45a). However, a person may ransom himself, his wife, or any other relative, for any amount [43], pp. 91-107). Thus, we clearly see that societal ethics differ from individual ethics ([38], Chapter 26; [43], pp. 91-107; [51]). Hence, society may decide on its priorities in distributing its resources, even when individuals may be harmed by these decisions.

However, in individual patient-physician relationships triage decisions do not necessarily comport with such societal priorities. When shortage of resources or time requires a triage decision on the level of micro-allocation, there are several halakhic rules that have been proposed by some decisors. First, if two sick people present at the same time, and lifesaving measures are available for only one of them, the one who has greater chance to survive takes precedence over the other ([18]; [56]). Second, if treatment of one was already initiated, it is forbidden to withdraw it from that person, even if he has little or no chance to survive, for the sake of any other patient ([5]; [19]). Third, if two people present at the same time, and their medical condition is equal, there is a Talmudic list of priorities, based on social worth and scholarship (*Tractate Horayot* 13a; Maimonides' *Mishneh Torah, Matnot aniyim* 8:17). Some decisors, however, have questioned the practical applicability of this list [23], or whether it applies to life-threatening situations [70]. Fourth, age is irrelevant to triage decisions [23]. Fifth, if one is the owner of the lifesaving resource and he needs it for his own survival, he takes precedence over any other person in need of that resource (*Babylonian Talmud, Tractate Baba Mezi'ah* 62a).

H. EUTHANASIA

The Halakhic literature concerning the various aspects of euthanasia is very extensive. In this review only the major attitudes will be presented.

The most significant factor in determining the approach towards the terminally ill patient is the concept that human life is of inestimable value and of paramount importance. Judaism regards every moment of life as sacred. Therefore, the Jew must even suspend religious observance for the higher reality, life itself. Every person is duty-bound to save and protect not only his own life and health but also that of his fellow human beings.

Every life is equally valuable, and there can be no distinction between one person and another. The requirement to save physically or mentally defective persons abrogates all other laws, in the same way as the saving of normal and healthy people. Hence, the physician's duty does not end when he is incapable of restoring the lost health of his patient ([10], Chapter 3).

Any form of active euthanasia is strictly prohibited and condemned as plain murder. According to Jewish law it is prohibited to perform any positive action that might directly and intentionally hasten death, even if death is hastened only by a matter of moments. Hence, anyone who kills a dying person is liable to the death penalty as a common murderer ([17]; [32]).

This accepted Halakhic position is based on several Talmudic statements concerning the dying person: "One who is in a dying condition is regarded as a living person in all respects. One may not close the eyes of the dying person. He who touches the dying or moves him is shedding blood" (*Tractate Semachot* 1:1-4; Maimonides' *Mishneh Torah*, *Avel* 4:5); "He who closes the eyes of a dying person while the soul is departing is a murderer. This may be compared to a lamp that is going out. If a man places his finger upon it, it is immediately extinguished" (*Babylonian Tractate Shabbat* 151b; Maimonides' *Mishneh Torah*, *Avel* 4:5).

Whether passive euthanasia is permissible or not is still debatable among contemporary Halakhic authorities. Some maintain that physicians and others are obligated to make use of any medical resources which are available in order to prolong the life of a terminally ill patient, even if the patient will only live for a brief period. Accordingly, withholding, and certainly withdrawing, life-sustaining therapies would be forbidden, and any action that results in hastening of the death of a dying patient would be considered an act of murder, even if the patient requests it ([10], pp. 134-145; [59]). However, even according to this viewpoint, the physician is not obligated to employ procedures which are themselves hazardous in nature and may potentially shorten the life of the patient. Nor is either the physician or the patient obligated to employ a therapy which is experimental in nature ([10], 134-145).

Other authorities sanction the position that not all available measures have to be applied if they merely prolong the dying process. In fact, when medical opinion is clear that the prognosis is very poor and the patient is suffering, one should withhold futile life-prolonging measures ([17]; [19]). However, the following distinction is made by these authorities: means that are natural, or means provided for treatable conditions unrelated to the terminal illness, must be continued to the very end. By contrast, futile treatments that are directly related to the disease which causes the terminal state can and should be withheld. Hence, oxygen, fluids, nutrition, antibiotics, and pain-relief medications must be maintained for a terminally ill patient until the very end, even against his will ([3]; [20]; [22]; [66]). On the other hand, chemotherapy, dialysis, or resuscitation, for instance, ought not be instituted when deemed futile. Also, at the patient's request, it is permissible to withhold medications and treatments which might cause him great pain and discomfort [3].

A terminally ill person may be given oral or parenteral narcotics to relieve his pain, even at the risk of depressing his respiratory center, provided the narcotics are prescribed solely for the purpose of relieving pain and not to hasten death ([6]; [17]; [62]).

I. DEFINITION OF THE MOMENT OF DEATH

The criteria for death under Jewish law is an issue that has been the subject of a heated controversy among contemporary Halakhic authorities. The classic definition of death is the total stoppage of the circulation of the blood, and the cessation of the animal and vital functions consequent thereon, such as respiration, pulsation, etc.

All Jewish authorities agree that cerebral or neocortical death, viz., a persistent vegetative state, is not death according to Jewish law, because respiration in this situation is preserved normally and spontaneously. Also, it is certainly accepted that the heart is vital to life, since life is not possible in its absence. However, this is true also for the liver, kidneys, or pancreas. The crucial issue is whether the activity of the heart constitutes an integral and necessary part of the Halakhic definition of death. It should be stressed that the metaphysical characteristics delegated to the heart in ancient Jewish sources (i.e., emotion, wisdom, feeling, will) are irrelevant to the Halakhic discussion. Hence, an intense debate is ongoing among rabbinic authorities as to whether or not Jewish law recognizes whole brain (i.e., cortical and brain-stem) death as a definition of death.

The classic definition of death in Judaism as found in the Talmud (*Babylonian Talmud*, *Tractate Yoma* 85a) and in the Codes of Jewish Law (Maimonides' *Mishneh Torah*, *Shabbat*, 219; Karo's *Shulchan Aruch*, *Orach Chaim* 329:4) is the irreversible absence of spontaneous respiration in a person who appears dead, viz., shows no movements and is unresponsive to all stimuli. The Talmudic discussion, which serves as the primary source for establishing the criteria of death in Jewish law (*Tractate Yoma* 85a), indicates that death may be assumed to coincide with cessation of respiration. The Talmud cites the verse "all in whose nostrils is the breath of the spirit of life" (*Genesis* 7:22) as indicating that life is present as long as respiratory activity persists. Also, in presenting the details concerning the creation of Adam, the Torah states, "and He blew into his nostrils the soul of life" (*Genesis* 2:6). The soul is thus described as having entered the body through the nostrils. The above mentioned Talmudic paragraph speaks of examining the nostrils as the most reliable criterion of death.

Thus, the Biblical verses and the Talmudic texts perceive breathing as a sign of life, and absence of breathing as a sign of death. It follows that death occurs when breathing totally and irreversibly ceases. According to current scientific knowledge, breathing is dependent upon the brainstem and not the heart. Accordingly, in order to prove that spontaneous breathing has irreversibly stopped, it is necessary convincingly to demonstrate either the irreversible cessation of the heartbeat or the irreversible cessation of the activity of the brainstem. When either of these situations exists, one can be certain that spontaneous breathing will not resume, and that the person is Halakhically dead. Therefore, the Halakhic definition of the moment of death – absolute and irreversible cessation of spontaneous respiration – was, and remains, the sole Halakhic definition. Only the technical means of ascertaining the moment of death have changed ([45]; [46]).

The Chief Rabbinate of Israel, on November 3, 1986, unanimously endorsed the acceptance of the permanent and irreversible cessation of breathing as the Halakhic definition of death. The verification of this situation should be done by proving that the brain, including the brainstem which controls spontaneous respiration, has been totally and irreversibly functionally destroyed.

A Halakhic variant of the concept of brain death is the concept of "physiologic decapitation" as an acceptable definition of death in Judaism, even if cardiac function has not ceased ([50]; [54]). This is based on the Talmudic deliberation concerning a person who has been decapitated. His heart may still function for a brief period of time, but that person is Halakhically dead at the moment the whole brain was severed from the remainder of the body. If one can medically establish that there is total and irreversible cessation of all brain functions, the patient is as if "physiologically decapitated".

Dissenting opinions among contemporary Rabbis do not accept brain death as the definition of death insofar as Jewish law is concerned, since the sole criterion of death according to these authorities is total cessation of both cardiac and respiratory activity. They suggest that cardiac activity is crucial in determining whether death has occurred ([12]; [69]).

J. ORGAN TRANSPLANTATION

The Halakhic issues concerning transplantation can be subdivided into those pertain to the recipient and those that pertain to the donor.

The major issue concerning the recipient is whether or not he is allowed to subject himself to the danger of the operative procedure. The question is whether the recipient transgresses the Biblical commandment "Take heed to thyself and keep thy soul diligently" (*Deuteronomy* 4:9), which is Halakhically interpreted to mean the removal or avoidance of all danger to one's physical well-being (*Babylonian Tractete Berachot* 32b; Maimonides' *Mishneh Torah*, *Rotze'ach* 11:4). Based on the current state of the art, in which greater than fifty-percent chances for survival for longer than one year post-transplantation is the norm for most organs, it is permissible to undergo all types of medically established organ transplantations.

From the Jewish standpoint it is preferable to use either artificial organs or organs derived from animals. At the moment, however, these options are technically unattainable. Hence, the only practical donors are human beings, living or dead.

The most significant question posed in consideration of whether or not organ transplants from a living donor are permissible is the justification of the hazard to which the donor is subjected in such procedures. The question is whether or not a person is obligated, or even permitted, to place his life in danger in order to preserve the life of his fellowman. Does the donor transgress the Biblical commandment "Take heed to thyself and keep thy soul diligently" (*Deuteronomy* 4:9)? Based on the *responsum* by R. David ibn Zimra (1479-1573) (*Responsa Radbaz*, Part III, No. 627), one who jeopardizes his own life is considered a "pious fool". According to ibn Zimra, it is inconceivable that the Torah, "whose paths are paths of pleasantness" (*Proverbs* 8:17), would demand such a great sacrifice of any person as a matter of obligation. However, when the potential danger is considered to be small, a person who saves the life of another has clearly performed an act of loving kindness of the highest order. The majority of rabbinic authorities rule that it is permissible, but not required, for a living donor to subject himself to the small risk involved in this procedure in order to save the life of another ([4]; [15]; [26], pp. 251-259; [29]; [31]). A prominent Halakhic authority has ruled that such a donation is obligatory according to Jewish law [75], whereas a minority of current rabbis is of the opinion that it is forbidden to donate organs because of the potential danger involved in the procedure for the donor ([52]; [57]).

Cadaver transplants and the permissibility of willing one's organs pose a different set of problems. The considerations with regard to these questions are closely akin to those posed by autopsies. The following issues are of Halakhic concern: the prohibition of desecrating or mutilating the dead, the prohibition of deriving benefit from the dead, the prohibition of delaying the burial of the dead, and the positive commandment of burying the dead. All these concerns and prohibitions are set aside if necessary in order to eliminate a danger to the life of a human being. Another Halakhic principle is that the permission to perform an act in order to save life which otherwise is prohibited is effective only if the person is at hand, "before us". However, according to many contemporary decisors, whenever there is a reasonable certainty that the organs will be eventually used in life-saving operations, it is permissible to harvest them and keep them in an "organ-bank".

In all cases, an organ or tissue may be removed only after the donor has been pronounced dead in accordance with the criteria of Jewish law. Thus, the moribund patient may not be moved, nor may other procedures be initiated, until death has actually taken place. The definition of the exact moment of death is still debatable among contemporary Jewish scholars, as was analyzed above concerning the definition of the moment of death.

Permission is necessary either from the deceased prior to his demise or from the next of kin ([15]; [73]), except for emergencies.

Hadassah Medical School
Hebrew University
Jerusalem, Israel

APPENDIX: BOOKS AND PERIODICALS DEVOTED ENTIRELY TO JEWISH MEDICAL LAW AND ETHICS

English Books (in chronological order)

Zimmels, H.J.: 1952, *Magicians, Theologians and Doctors*, Edward Goldston & Son, London.

Feldman, D.M.: 1968, *Marital Relations, Birth Control and Abortion in Jewish Law*, Schocken, New York.

Jakobovits, I.: 1975, *Jewish Medical Ethics*, 2nd ed., Bloch Publishing Company, New York.

Rosner, F., and J.D. Bleich (eds.): 1979, *Jewish Bioethics*, Sanhedrin Press, New York.

Abraham, S.A.: 1980, *Medical Halacha for Everyone*, Feldheim Publishers, Jerusalem.

Steinberg, A.: 1980, *Jewish Medical Law*, Gefen Publishing, Jerusalem.

Bleich, J.D.: 1981, *Judaism and Healing*, Ktav Publishing House, Hoboken, New Jersey.

Rosner, F.: 1984, *Medicine in the Mishneh Torah of Maimonides*, Ktav Publishing House, Hoboken, New Jersey.

Steinberg, A. (ed.): 1989, *The European Colloquium on Medical Ethics: Jewish Perspectives*, Magnes Press, Hebrew University, Jerusalem.

Rosner, F.: 1990, *Modern Medicine and Jewish Ethics*, 2nd revised ed., Ktav Publishing House, Hoboken, New Jersey.

Hebrew Books (in chronological order)

Perlman, M.: 1926-1934, *Midrash Ha'Refuah*, 3 vols., Dvir Publishing, Tel Aviv.

Abraham, S.A.: 1976-1978, *Lev Avraham*, 2 vols., Feldheim Publishers, Jerusalem.

Steinberg, A.: 1978, *Hilchot Rofim U'Refuah*, Mossad Harav Kook, Jerusalem.

Abraham, S.A.:, 1983-1986, *Nishmat Avraham*, 3 vols., private publication, Jerusalem.

Zilberstein, Y., and M. Rothshield: 1987, *Torat Ha'Yoledet*, 2nd ed., private publication, Bnei Braq.

Steinberg, A.: 1988, *Encyclopedia Hilchatit Refuit*, 2nd ed., Vol. 1, Dr. Falk Schlesinger Institute for Medical Halakhic Research at Shaare Zedek Medical Center, Jerusalem.

Hebrew Periodicals (in chronological order)

Assia (quarterly), Dr. Falk Schlesinger Institute for Medical Halakhic Research at Shaare Zedek Medical Center, Jerusalem, since 1970.

Assia Anthology (irregular), Dr. Falk Schlesinger Institute for Medical Halakhic Research at Shaare Zedek Medical Center, Jerusalem, since 1976.

Be'Shvilei Ha'Refuah (irregular), Laniado Hospital, Netania, since 1979.

Halakha U'Refuah (annually), Regensberg Institute, Chicago-Jerusalem, since

1980.

Ha'Refuah Le'Or Ha'Halakhah (irregular), Institute for Research in Medicine and Halakhah, Jerusalem, since 1980.

Emek Halakhah (irregular), published privately, Jerusalem, since 1986.

BIBLIOGRAPHY

1. Abraham, S.A.: 1985, *Nishmat Avraham*, *Yore De'ah* 155:2, Jerusalem (Hebrew).
2. Abraham, S.A.: 1989, "A Baby With a Severe Defect", in: A. Steinberg (ed.), *Emek Halakhah*, Vol. 2, Yeshiva University Press, New York (Hebrew).
3. Auerbach, S.Z.: 1986, *Responsa Minchat Shlomo*, No. 91, Sec. 24, Jerusalem (Hebrew).
4. Auerbach, S.Z.: cited in Abraham, S.A.: 1985, *Nishmat Avraham*, *Yore De'ah* 157:4(2) (Hebrew).
5. Auerbach, S.Z., cited in Abraham, S.A.: 1985, *Nishmat Avraham*, *Yore De'ah* 252:2, Jerusalem (Hebrew).
6. Auerbach, S.Z., cited in Abraham, S.A.: 1985, *Nishmat Avraham*, *Yore De'ah* 339:4, Jerusalem (Hebrew).
7. Auerbach, S.Z., cited in Abraham, S.A.: 1987, *Nishmat Avraham*, *Choshen Mishpat*, No. 425:1(15), Jerusalem (Hebrew).
8. Auerbach, S.Z., cited in Abraham, S.A.: 1987, *Nishmat Avraham*, *Choshen Mishpat*, No. 425:1(21), Jerusalem (Hebrew).
9. Bakshi-Doron, E.: 1982, *Responsa Binyan Av*, Vol. 1, No. 50 (Hebrew).
10. Bleich, J.D.: 1981, *Judaism and Healing*, Ktav Publishing House, Hoboken, New Jersey.
11. Bleich, J.D.: 1979, "Experimentation on Human Subjects", in F. Rosner and J.D. Bleich (eds.), *Jewish Bioethics*, Sanhedrin Press, New York.
12. Bleich J.D.: 1989, "Of Cerebral, Respiratory and Cardiac Death", *Tradition* 24(3), 44-66.
13. Breisch, M.Y.: 1966, *Responsa Chelkat Yaakov*, Vol. 3, No. 136 (Hebrew).
14. *Encyclopedia Judaica*: 1972, "Halakhah", Vol. 7, Keter Publishing House, Jerusalem.
15. Feinstein, M.: 1973, *Responsa Iggrot Moshe*, *Yore De'ah*, Vol. 2, No. 174, Sec. 4, New York (Hebrew).
16. Feinstein, M.: 1985, *Responsa Iggrot Moshe*, *Choshen Mishpat*, Vol. 2, No. 69, Bnei Braq (Hebrew).
17. Feinstein, M.: 1985, *Responsa Iggrot Moshe*, *Choshen Mishpat*, Vol. 2, No. 73, Sec. 1, Bnei Braq (Hebrew).
18. Feinstein, M.: 1985, *Responsa Iggrot Moshe*, *Choshen Mishpat*, Vol. 2, No. 73, Sec. 2, Bnei Braq (Hebrew).
19. Feinstein, M.: 1985, *Responsa Iggrot Moshe*, *Choshen Mishpat* Vol. 2, No.

74, Sec. 1, Bnei Braq (Hebrew).
20. Feinstein, M.: 1985, *Responsa Iggrot Moshe*, *Choshen Mishpat*, Vol. 2, No. 74, Sec. 3, Bnei Braq (Hebrew).
21. Feinstein, M.: 1985, *Responsa Iggrot Moshe*, *Choshen Mishpat*, Vol. 2, No. 75, Sec. 2, Bnei Braq (Hebrew).
22. Feinstein, M.: 1985, *Responsa Iggrot Moshe*, *Choshen Mishpat*, Vol. 2, No. 75, Sec. 4, Bnei Braq (Hebrew).
23. Feinstein, M.: 1985, *Responsa Iggrot Moshe*, *Choshen Mishpat*, Vol. 2, No. 75, Sec. 7, Bnei Braq (Hebrew).
24. Feldman, D.M.: 1974, *Marital Relations, Birth Control and Abortion in Jewish Law*, Schocken Books, New York.
25. Grosness, A.L.: *Responsa Lev Arie*, Vol. 2, No. 32 (Hebrew).
26. Halevi, C.D.: 1983, "Organ Transplantation According to Halakha", in A. Steinberg (ed.), *Assia*, Vol. 4, Reuven Mass Publication, Jerusalem (Hebrew).
27. Halevi, C.D.: 1989, "Reduction of Fetuses", *Assia* 47-48, 14-17 (Hebrew).
28. Hershler, M.: 1980, "Halakhic Issues Concerning a Test-Tube Baby", *Halakha U'Refuah* 1, 307-320 (Hebrew).
29. Hershler, M.: 1981, "Kidney Donation by an Incompetent Person", *Halakha U'Refuah* 2, 122-128 (Hebrew).
30. Israeli, S.: *Responsa Amud Ha'Yemini*, No. 32 (Hebrew).
31. Israeli, S.: 1986, "Editorial Remarks", *Barkai* 3, 35-36 (Hebrew).
32. Jakobovits, I.: 1961, "The Dying and Their Treatment in Jewish Law: Preparation for Death and Euthanasia", *Hebrew Medical Journal* 2, 251 ff.
33. Jakobovits, I.: 1959, 1975, *Jewish Medical Ethics*, Bloch, New York.
34. Jakobovits, I.: 1988, "Some Modern Responsa on Medico-Moral Problems", *Assia: Jewish Medical Ethics* 1, 5-16.
35. Kook, S.: 1983, "Medical Experimentation", in A. Steinberg (ed.), *Assia*, Vol. 3, Reuven Mass Publication, Jerusalem (Hebrew).
36. Nebenzhal, A.: 1986, "In Vitro Fertilisation – Remarks", in M. Halperin (ed.), *Assia*, Vol. 5, Reuven Mass Publishing, Jerusalem (Hebrew).
37. Rafael S.: 1983, "Medical Confidentiality – Halakhic Aspects", in A. Steinberg (ed.), *Assia*, Vol. 3, Reuven Mass Publication, Jerusalem (Hebrew).
38. Rosner F.: 1990, *Modern Medicine and Jewish Ethics*, 2nd revised ed., Ktav Publishing House, Hoboken, New Jersey.
39. Silberstein, Y., cited in Abraham, S.A.: 1987, *Nishmat Avraham*, *Choshen Mishpat*, No. 425:1(19), Jerusalem (Hebrew).
40. Silberstein, Y.: 1989, "Reduction of Fetuses", *Assia* 45-46, 62-68 (Hebrew).
41. Silberstein, Y.: 1989, "Resuscitation of a Baby Born Less Than Six Months of Gestation", *Bishvilei Ha'Refuah* 9, 81-85 (Hebrew).
42. Steinberg A.: 1986, "The Treatment of a Defective Newborn", *Techumin*

7, 226-230 (Hebrew).

43. Steinberg A.: 1989, "Distribution of Scarce Medical Resources in Medicine – Ethical and Halakhic Aspects", in A. Steinberg (ed.), *Emek Halakhah*, Vol 2, Yeshiva University Press, New York (Hebrew).
44. Steinberg, A.: 1988, *Encyclopedia Hilchatit Refuit*, 2nd ed., Vol. 1, The Dr. Falk Schlesinger Institute for Medical Halakhic Research at Shaare Zedek Medical Center, Jerusalem (Hebrew).
45. Steinberg, A.: 1988, "The Determination of the Moment of Death and Heart Transplantation", *Assia* 44, 56-77 (Hebrew).
46. Steinberg, A.: 1988, "The Determination of the Moment of Death and Heart Transplantation", *Or Ha'Mizrach* 36(1), 48-65 (Hebrew).
47. Steinberg, A.: 1989, "Bioethics: Secular Philosophy, Jewish Law, and Modern Medicine", *Israeli Journal of Medical Science* 25, 404-409.
48. Steinberg, A.: 1991, *Encyclopedia Hilchatit Refuit*, Vol. 2, The Dr. Falk Schlesinger Institute for Medical Halakhic Research at Shaare Zedek Medical Center, Jerusalem (Hebrew).
49. Sternbuch, M.: 1987, "Test-Tube Baby", *Bi'Shvilei Ha'Refuah* 8, 29-36 (Hebrew).
50. Tendler, M.D.: 1978, "Cessation of Brain Function. Ethical Implications in Terminal Care and Organ Transplants", *Annals of the New York Academy of Sciences* 315, 394-497.
51. Tendler, M.D.: 1984, "Rabbinic Comment: Triage of Resources", *Mount Sinai Journal of Medicine* 51, 106-109.
52. Toledano, P.: 1986, "Concerning Kidney Transplantation", *Barkai* 3, 23-36 (Hebrew).
54. Veith, F.J., *et al*: 1977, "Brain Death. I. A Status Report of Medical and Ethical Considerations", *Journal of the American Medical Association* 238, 1651-1655.
55. Waldenberg, E.Y.: 1965, *Responsa Tzitz Eliezer*, Vol. 8, No. 36, Jerusalem (Hebrew).
56. Waldenberg, E.Y.: 1967, *Responsa Tzitz Eliezer*, Vol. 9, No. 17, Sec. 5, Jerusalem (Hebrew).
57. Waldenberg, E.Y.: 1967, *Responsa Tzitz Eliezer*, Vol. 9, No. 45, Jerusalem (Hebrew).
58. Waldenberg, E.Y.: 1967, *Responsa Tzitz Eliezer*, Vol. 9, No. 51, Jerusalem (Hebrew).
59. Waldenberg, E.Y.: 1970, *Responsa Tzitz Eliezer*, Vol. 10, No. 25 Sec. 6, Jerusalem (Hebrew).
60. Waldenberg, E.Y.: 1973, *Responsa Tzitz Eliezer*, Vol. 11, No. 45, Sec. 13, Jerusalem (Hebrew).
61. Waldenberg, E.Y.: 1978, *Responsa Tzitz Eliezer*, Vol. 13, No. 81, Jerusalem (Hebrew).
62. Waldenberg, E.Y.: 1978, *Responsa Tzitz Eliezer*, Vol. 13, No. 87, Jerusalem (Hebrew).

63. Waldenberg, E.Y.: 1978, *Responsa Tzitz Eliezer*, Vol. 13, No. 88, Jerusalem (Hebrew).
64. Waldenberg, E.Y.: 1978, *Responsa Tzitz Eliezer*, Vol. 13, No. 102, Jerusalem (Hebrew).
65. Waldenberg, E.Y.: 1981, *Responsa Tzitz Eliezer*, Vol. 14, No. 27, Jerusalem (Hebrew).
66. Waldenberg, E.Y.: 1981, *Responsa Tzitz Eliezer*, Vol. 14, No. 80, Jerusalem (Hebrew).
67. Waldenberg, E.Y.: 1983, *Responsa Tzitz Eliezer*, Vol. 15, No. 45, Jerusalem (Hebrew).
68. Waldenberg, E.Y.: 1985, *Responsa Tzitz Eliezer*, Vol. 16, No. 4, Jerusalem (Hebrew).
69. Waldenberg, E.Y.: 1988, *Responsa Tzitz Eliezer*, Vol. 17, No. 66, Jerusalem (Hebrew).
70. Waldenberg, E.Y.: 1990, *Responsa Tzitz Eliezer*, Vol. 18, No. 1, Jerusalem (Hebrew).
71. Waldenberg, E.Y.: 1990, *Responsa Tzitz Eliezer*, Vol. 18, No. 19, Jerusalem (Hebrew).
72. Weiss, A.H.: 1924, *Dor Dor Ve'Dorshav*, Vol. 2, New York & Berlin.
73. Weiss, Y.Y.: 1975, *Responsa Minchat Yitzchak*, Vol. 5, No. 7, Jerusalem (Hebrew).
74. Yoseph, O.: *Responsa Yabi'ah Omer*, Vol. 4, *Even Ha'Ezer* No. 1, Jerusalem (Hebrew).
75. Yoseph, O.: 1980, *Responsa Yechave Da'at*, Vol. 3, No. 84, Jerusalem (Hebrew).
76. Yoseph, O.: 1981, *Responsa Yechave Da'at*, Vol. 4, No. 60, Jerusalem (Hebrew).
77. Yoseph, O., cited in: Abraham, S.A.: 1987, *Nishmat Avraham*, *Even Ha'Ezer* 1:5, Jerusalem (Hebrew).
78. Zager, M.Z.: 1988, *Responsa Va'Yashev Moshe*, Vol. 1, No. 74 (Hebrew).

KENNETH L. VAUX

BIOMEDICAL ETHICS IN THE REFORMED TRADITION

Like most religious traditions, the Reformed movement (churches stemming from Calvin's reformation) is today a pluralistic and dynamic phenomenon. All attempts, therefore, to quick-freeze a slice which would somehow represent both a norm and the present state of thought inevitably end up violating the inherent variety and vitality which is the ongoing character of any tradition *semper reformando*. In addition to this difficulty, one is never quite sure where to look for that elusive "normative center" of a given tradition. The convictions and commitments of the laity are shown by all studies to differ dramatically from those of the clergy. In Presbyterian assemblies, for example, lay, and even clerical, views are often even more discordant with the views of denominational leaders and appointed working groups. Does one look to the teaching and writing theologians who may represent cutting-edge thought that has not yet penetrated the ecclesiastical and provincial realms of the Church's life?

To represent the biomedical ethics of the Reformed tradition at the turn of the last decade of the twentieth century will require that we consult both official documents (aware that they are almost always misrepresentative of the mind of the Church), and the more creative, albeit avant garde, thought of the theologians. Following some background material, I will delineate the topic by reviewing the General Assembly Minutes of the United Presbyterian Church in the U.S.A. during the middle and late 1980s on biomedical themes. As a primary author of many of those positions and policies, I will then interpret and interpolate my own bioethics. At the same time, I will extract moral teachings on some themes from the conservative branch of the Reformed movement, using an important document of a working group at Calvin College [3].

BACKGROUND

Reformed churches throughout the world claim a constituency of nearly 40 million. To the old Reformed Churches of Europe (Switzerland, France, Germany, Holland, Hungary, Czechoslovakia, Scotland, England, etc.) are added the mission churches in Asia, Africa, and South America, e.g. South Korea, and the colonial churches in North America. These latter communions, the Reformed and Presbyterian Churches of the U.S.A. and Canada, will be the focus of this report.

The Calvinist churches of Europe and the New World bear the stigmata of the Reformation with its two-fold emphasis on radical obedience to the divine word and will, and on liberty, refusing to bend the conscience to any other authority, be it ecclesial, political, legal, or societal. Theologically focused convictions about life and health, birth, suffering, and death are thus joined to an affirmation of free and theonomous conscience on all matters of life and death. For this reason, in most Reformed bodies, ethical policy across the years has accented stewardship, communal and familial responsibility, and free choice when it comes to matters such as contraception, abortion, health care provision, and euthanasia.

As new challenges are met, all living religious and ethical traditions splay into conservative and progressive movements. Fundamentalism and liberalism are necessary antinomies in any vital faith and work (ethical mission). On abortion, for example, the Reformed tradition will display its "holiness" heritage in a "sanctity of life" conviction embodied in fellowships like the Orthodox Presbyterian Church, as well as in its broad conservative constituency. Liberty of conscience and freedom of choice will be affirmed by the Hungarian Reformed Church and by the broad progressive constituency of the worldwide communion. Euthanasia and physician-assisted suicide will be censured by the Christian Reformed Church in Michigan and cautiously endorsed by the Reformed denomination back in the mother country, Holland. Retaining Catholic substance and Protestant principle, to use Paul Tillich's helpful phrase, keeps both memory and hope vital, as responsibility requires. Religion (*religare*) means to be tethered (ligatured) to some frame of meaning and some possibility. All ethics is posited on the basis of that same discrepancy between an *is* and an *ought*.

As with their French, German, Scottish, and Puritan forebears, North American Presbyterians have always looked on the human body with awe and special trust, on sexuality in terms of accountability, on procreation as divine gift and sacred imperative, on the provision of education, shelter, food, and health as obligatory social policy arising from our own received blessing, on life as a gift to relish and relinquish in service, and on death as life's culmination to be received in due season in grace and peace.

THEOLOGICAL MOTIFS IN REFORMED BIOETHICS

This religious heritage and present faith constitutes a powerful source of beliefs and values about health matters both for individual and society. The sovereignty of God, the reality of sin and forgiveness, the grace of Christ, the gift and responsibility of life, the imperative to love and justice, and the Lordship of God over nature, history, and human destiny are themes which animate belief and value in the Reformed Faith.

As the creating, sustaining and consummating Being, God gives life, safeguards well-being, and receives death. Recognition of this dependence yields the dispositions of respect, stewardship, and ultimate acquiescence to God's will as we live our lives, generate families, experience illness, and face death. Redeeming purpose and providence are qualities of God's character which imprint our existence. To succeed in life is to discern and accede to that plan and good will. Virtue or health is to advance that kingdom in all of one's action. In concerns of health and medicine, sovereignty doctrine has generated throughout history a curious blend of scientific and technical efforts to ameliorate the human condition and its ills, and serene resignation in the face of deleterious forces, all the while confidence in an overriding divine purpose.

The Puritan ethic, for example, a cultural phenomenon in part created by Reformed Faith, first articulated and propelled the medical value of prolonging life. At the same time it developed a most exquisite theology of anticipation, even celebration, of death.

The Lordship of God places parental and regal expectation on His children, which expectation they have forsaken, and from which possibility they have fallen short. Violating our designed nature and intended destiny, we have sullied our native well-being. Sin allows disease to fester in our being, alienation to scar our emotional and relational lives, and death to cloud the horizon of our future. The drama of sin and grace punctuates our being most poignantly in disease and health.

"My only comfort in life and death...my faithful savior, Jesus Christ" (*Heidelberg Catechism*). With these words a second motif of the Reformed Faith bearing on health and disease is expressed. The grace of Christ overwhelms our boundedness to sin and death, and renders possible the righteousness or health (salvation) of God. In Christ we are heirs to the grace of life, that is, to health, love, marriage, progeny, longevity, and the good conscience of His service.

These existential doctrines of creation, fall, and redemption, especially given the tenor they develop within Calvinist history, activate a high sense of responsibility for life, health, and general welfare. Along with the Jewish, Catholic, Lutheran, and Anglican communions, Presbyterians and Baptists found hospitals, medical schools, and clinics, and send sons and daughters into the ministries of physical, mental, and spiritual health. Commitment is also inspired within this community to social justice as equity, and concern for the disadvantaged and charity are encouraged.

Finally, the Reformed faith proclaims God as the beginning, center, and end of our existence. We die, not to oblivion, but to God's presence. Death is the announcement that our work is done, that we have completed our earthly ministry. Though fighting disease and premature death is noble, our faith asks us to receive death as if entering a new life established by God who is faithful. Like Judaism, Reformed Christians honor life, health, the body, and longevity. In Christ we also receive pain, suffering, and death as the Lord's chastising, cleansing, and culminating will. These underlying beliefs give rise to applied ethics in the following areas.

I. NEW REPRODUCTIVE TECHNIQUES

The Reformed tradition would cringe at the heading. The deep and rich privilege and opportunity of procreation has now been transmuted into the 18th-century mechanical metaphor of reproduction. One might ask, Will we soon shift the metaphor again to Huxley's "decantation"? Affirming the mystery and nobility of receiving and transmitting the lives of new persons into God's world has always been an emphasis of Reformed thought. At the same time, the tradition has celebrated science and technology as applied to human generativity. This is especially true in its Renaissance European (England, Scotland, Holland), Puritan (England and New England), and modern American ethos.

By way of background to Church ethics on this issue, note must be made on what has been the most consuming bioethical issue to face the Reformed communion, that of the ethics of sexuality, especially homosexuality. In 1976, and again in 1988, the denomination appointed expert commissions, including scientists, social scientists, physicians, ethicists, and laypersons, to formulate a moral appraisal of the condition and practice of homosexuality, and to give guidance on the even more vexing question of ordination of avowed, practicing homosexuals to ministry. The General Assembly of the Church concluded in 1978 that biblical and theological teaching precluded the ordination of avowed, practicing homosexual persons to Gospel ministry. It also affirmed the full civil rights of such persons, and advocated ministries of understanding and compassion. In an unusual action, the General Assembly (1978) adopted a policy which grew out of a minority report of the working task force.

> [T]he minority recommends that the General Assembly exercise its judicial role as the highest judicatory of the church...and determine a definitive interpretation of the constitution which specifies that self-affirming, practicing homosexual persons may not be ordained to the professional ministry, or to the offices of ruling elder and deacon.

> Since homosexuality is one issue that helps clarify our general responsibility to God in his world, and focuses many dimensions of belief and action, such activities should probe such basic issues as: (1) the strengthening of family life; (2) nurturing lifestyles in our families, congregations, and communities that celebrate the values of friendship with peers of one's own sex and the opposite sex, committed choice of life-mates, joyous and loving fidelity within marriage, the establishment of homes where love and care can build strong children able to give loving service to others, and the fashioning of an atmosphere of justice, truth, and kindness that signals Christ's presence; (3) understanding how to extend ministry of deep concern and challenge to those who through choice or circumstance are sexually active, homosexually or heterosexually, outside the covenant of marriage; (4) helping those whose joy and tenderness is destroyed by homophobia, the irrational fear of and contempt for homosexual persons ([2], pp. D-198-199).

Teaching on homosexuality is part of the fabric of the Church's thought about sexuality, parenting, and family life, all of which forms the foundation of ethics across the life-cycle. On the wider range of new technological possibilities surrounding procreation, the Church affirms the use of contraception, supports interventions to enhance fertility so that those who could not have children might be able to do so, encourages adoption, and warns against all forms of inequity and injustice in the provision of the new technologies. The following passages capture the thrust of Church teaching at this point:

> ...[W]hile the option of bearing children should be available as universally as possible, to bear a child should not be undertaken without clear intentionality...[The Church] affirms the use of therapies to overcome anovulation and other problems of infertility...affirms artificial insemination and in vitro fertilization...urges couples who cannot conceive to consider adoption...and urges further study on surrogacy ([1], p. 364).

One sees in these documents a progenerative theological bias, a cautious and critical affirmation of new technologies to support this good, and a firm demand for nonexploitation and justice.

In the area of genetic medicine and fetal diagnosis, Church policy emphasizes a blend of gracious acceptance of the lives which God's hand has extended to us, however malformed or "abnormal". It extols the careful and just utilization of biotechnology to safeguard lives and enhance the possibility of fulfillment of persons. It warns against manias toward perfection and the creation of "superhumans".

> ...In the area of medical genetics, priority should be given to the prevention of disease via family planning, genetic counselling and fetal diagnosis...([1], pp. 364-365).

Bouma *et al.* summarize the broad conserving tendency of Reformed faith on these matters. By quoting Oliver O'Donovan, the Anglican moralist, who argued in terms of the Nicene creed that children are "begotten not made", the report holds that children are gifts to be received in stewardship, not fabrications to afflict with our terrible and capricious will. The neocalvinist version of providence doctrine and divine ordination/destination of persons' lives takes the Augustinian confidence in God's handiwork in natural and historical process and adds to it the spiritual insight of the new (Reformation, Puritan) view of science and human freedom and power to fashion a perinatal philosophy strongly in concurrence with nature in light of human limitation, yet committed to transcendence over nature in service of human good and justice. Necessity and freedom are the two marks of conscience in the Reformed faith.

II. ABORTION

Moral doctrine on abortion intensifies this antinomy of freedom and necessity in the ethics of the Reformed tradition. The benign goodness of nature expressed in Catholic teleology and other natural law conviction (Thomism and early Calvinism) is now set aside, and the caprice and bondage of nature is emphasized so as to incline the moral equation toward dominion and freedom.

Freedom of choice thus comes to prevail over right to life in the tradition. Regrettably, to this author at least, a simplistic "prochoice" philosophy – univocal and unnuanced – now becomes the ideological counterpoint of the "right to life" harangue. Libertarian conscience has therefore come to dominate the abortion ethics of UPUSA (The United Presbyterian Church in the U.S.A.). After explicating and documenting this position, we will take note of two other moral options still very lively within the tradition.

The prevailing pro-abortion position is reasoned like this:

- Current contraception technology is far from reliable.
- We must respond sympathetically to various kinds of unintended pregnancies: a genetic problem, rape or incest, failed contraception, teenage pregnancies, felt limits to human resources.
- Abortion can be seen as a responsible choice within Christian responsibility.
- Public policy on abortion should reflect "religious liberty" and the decision for abortion must remain "with the individual".
- The unfolding liberation of women and the human community more generally entails freedom from the bondage created by unwanted pregnancy. ([1], pp. 367 ff.).

A more conservative version of a reformed ethic is found within the Christian Reformed Church and reflected in the Bouma report:

> ...Christians will normally perceive unborn human beings as precious gifts in the process of being welcomed into the covenantal community.
>
> Abortion often amounts to the refusal to make significant sacrifices to give a gift to someone else. We endorse a pro-life feminist view of sexuality...The abortions we would cooperate with or even recommend are what are often called the "hard cases" which probably amount to only a small percentage of abortions actually performed. ([3], pp. 223, 226).

A mediating view, which might be called a "sanctity of maternal/fetal life view", is expressed in the original working draft that I prepared for the UPUSA task force. Many overtures each year to General Assembly might suggest that this view is close to the sensus fidelium of our Church:

> The Presbyterian ethical heritage inclines toward an ambivalent imperative of a modified "Sanctity of Life" view of abortion. In our cultural development...human freedom is enlarged over natural necessity in the realm of sexuality and the procreative life....
>
> But surely we cannot see abortion as a feature of this pattern of liberation. Abortion is always tragic. For the woman it is the rude tearing of precariously conceived life from its sustaining dwelling, a profound physical and psychic trauma. For society, it is the precedent of life-sacrifice and the foreclosure of a potentially redeeming future. Abortion is not good. It is an evil but a mitigated evil....We need...to distinguish among differing situations and admit that there are times when the technique of abortion must be used. In the public policy arena we need to develop an ethic of reason and prudence. Then in a nuanced way we can assign value or disvalue to a given case according to some balance of the principles of beneficence, non-maleficence, justice and freedom.

A suggestive grouping of cases might be these: abortion would be strongly advised in incest, child rape, and cases where profound genetic defects (e.g. TaySachs, Lesch-Nayan Syndrome) are present or the pregnancy or delivery is lifethreatening. Abortion is permissible but not obligatory in coercive out-of-wedlock pregnancy, situations where pregnancy offers a physical or mental risk to the mother, moderately severe genetic or congenital abnormalities discovered in the fetus (Siamese twins, spina bifida, Down's Syndrome, etc.). Abortion is permissible but discouraged in illegitimate pregnancies, mild and treatable genetic or congenital accidents (hare lip, cleft palate, and the like). Abortion is proscribed for reasons of convenience, population control, and sex selection. Who should be the judge to say whether any of these conditions exist? The answer is obvious. The final choice should lie with the mother and father, together with medical and pastoral counsel. The law should convey general contours but abortion in any case should not be seen as a criminal act....

This proposal suggests that neither "right to life" nor "freedom of choice" doctrines satisfy. Both abandon the subtleties of freedom and conscience. Our society appears to be groping for some nuanced abortion policy that will more appropriately honor our cultural values of "sanctity of life", "freedom", and "justice". We must reject the morally simplistic reduction of this complex of issues into the "right to life", "life begins at conception" configuration on the one hand, and the equally absurd "freedom of choice" doctrine on the other. We need some carefully nuanced public policy which avoids the dangers of being restrictive or conducive for the tragic interim until completely reliable and trustworthy modes of contraception are available and used.

Abortion is a crisis for each of us but it is an especially difficult and prevalent problem for young (15-25) unmarried women who become pregnant. Statistics show that in 1988, eighty percent of those who sought abortion were unmarried. Thirty percent of the total 1.4 million were under twenty years of age and sixty-five percent were under twenty-five. In other words, nearly one million abortions were performed in 1988 for young, unmarried women for the main presenting cause of undesired pregnancy. The principal moral phenomenon we are dealing with, therefore, is that of a great proportion (perhaps ten percent) of our young adult female population finding themselves in an unacceptable trap from which they seek release. These women are not proud or particularly guilty; they simply find themselves in trouble and in need of help. Before condemnation or moral exhortation, the Church must extend mercy. Emulating Christ who drew near to the lost and the broken, the Church should be in the trenches with those who suffer rather than standing aloof in contempt and righteous indignation. The pilgrimage of grace is a story of repentance, restitution, and regeneration. The mercy of God and the grace of Christ come in at the extreme boundaries of our evil and estrangement. Our Church and society must abide in and offer that grace ([6], pp. 2 ff.).

III. MATERNAL/FETAL CONFLICT, CARE OF SEVERELY HANDICAPPED NEWBORNS

The recent bioethical issues of maternal/fetal conflict have yet to be addressed in any formal way by Reformed Church bodies. Given the bias of the tradition in favor of the mother and for freedom, pastoral moral guidance and policy might be expected to weigh against coerced obstetrical interventions to protect a fetus, such as doctor-ordered Caesarean section to deliver triplets, or house-detention for a pregnant cocaine-abusing mother, again to safeguard the well-birth of the fetus. The ethic of community care might seek pastoral programs to protect babies from neglect and need in areas of cocaine addiction and AIDS, for example.

On the theme of selective nontreatment of handicapped newborns, one can find the full spectrum of views among Presbyterians. On the one hand, former Surgeon General C. Everett Koop, an elder in the denomination, has been a forceful advocate for saving even severely malformed or diseased newborns. His perspectives on the subject have been in accord with his pronatalist films produced with conservative Swiss Reformed pastor, Frances Schaeffer. On the other hand, moderate and liberal bioethicists in the tradition, such as William May and Kenneth Vaux, have presented arguments to justify "letting die" in some circumstances.

Some of the motifs of the denominational teaching are found in the document *The Covenant of Life and the Caring Community*. It urges a team approach to decisionmaking in cases of neonatal crisis, in which the family, the physician(s), clergy, nurse, and social worker participate in the determination of the extent of treatment. It encourages consideration of all consequences of any decision to initiate lifesaving treatment that would cause more harm than benefit in the lives of extremely premature infants or those born with massive health problems ([1], p. 363). Obviously, the sanctity of life value, though central to the moral system, does not stand alone. Rather, a balance among respect for life, parental freedom, and a utilitarian rendering is consistently sought in order to yield the best ratio of beneficial outcomes to harm and costs.

IV. CONSENT TO TREATMENT, EXPERIMENTATION, AND CONFIDENTIALITY

The Advisory Council on Church and Society spoke to these issues as follows:

> The goals and practice of science shape its impact upon each person and profoundly affect the values of human culture. Research can be guided by the desire for profit and fame as well as by compassion and a sense of justice. Biomedical advances can be harnessed in the service of war or in the service of a life-giving society. Thus the 195th General Assembly (1983) calls upon scientists, institutions of scientific research and instruction, Congress, and all Presbyterians to advocate that:
>
> a. Research and development in science be guided by the following human values: survival, enhancement of life, justice and equity in access.
> b. Research endeavors involving the natural world be undertaken only with ecological consideration and care; that animal subjects be treated humanely; that human subjects be given the strongest human protections, including full information about the research, and that their consent be obtained without coercion; that fetal and embryonic research be undertaken with caution and sensitivity.
> c. Science place priority on supporting the basic life possibilities and needs of everyone rather than enhancing (prolonging, beautifying, etc.) the lives of a few, and that research and development allocations be made so as to reflect these goals.
> d. Biomedical Policy be formulated with participation by centers of value exposition, including the church ([1], p. 365).

In these pronouncements and recommendations, we see the prejudice of the tradition toward social justice and community values over private needs and desires. Though emphasis is given to human rights, confidentiality, informed consent, and patient participation in all forms of medical treatment and research, one finds here an unmistakable commitment to human protection, equitable access, and ecological sensitivity. In general, scientific progress is always to be chastened and challenged by human political purpose.

V. EQUITABLE ACCESS TO HEALTH CARE: ETHICAL ISSUES POSED BY COST CONTAINMENT MEASURES

Social justice and benevolent care for the poor and needy have always been pivotal values in Reformed ethics. The reformed tradition since the days of John Calvin in Geneva has been concerned for social transformation to serve the welfare of the community. The Puritans in England and America, chastened by a reformist faith and ethic, organized society to serve the public goods of education, housing, health, and general welfare. Taking Calvin's idea, which Karl Marx eventually cribbed, that "God wills equality among us so that none should have too little and none too much" (*Commentary on II Corinthians*), the Puritans sought to fashion an ideal commonwealth of concern for everyone's health, preservation of well-being and prevention of disease, succor and consolation to all who were sick, and gratitude to a gracious Healing Lord (*Christus Soter*).

In this light it is appropriate that the major bioethical concentration of Reformed communions at the turn of the ninth decade of this century is social justice, affordable access to health care, and the correlated concerns of health care for the poor, especially those who are young and old. These concerns are reflected in the 1988 document "Life Abundant: Values, Choices and Health Care". The Book of Order of the Presbyterian Church sets forth the leitmotif:

> healing and reconciling and binding up wounds, ministering to the poor and sick, the lonely, and the powerless, engaging in the struggle to free people from sin, fear, oppression, hunger and injustice, giving of itself and its substance to those who suffer sharing with Christ in the establishing of his just, peaceable, and loving rule in the world (Book of Order, G-3.0300) ([4], p. 517).

Here again the themes of freedom and bondage, realities derivative of grace and sin, frame the issue. Moral responses evoked by these realities are personal responsibility for health, and societal provision where freedom, opportunity, and ability to care for oneself fail. The recommendations reflect this carefully crafted tension:

> Desiring to give effective expression to the basic values of compassion, caring love, community wholeness and well-being, and justice that we hold to be fundamental in understanding and addressing the health issues and crises that confront the church and the nation, the 200th General Assembly (1988) of the Presbyterian Church (U.S.A.) therefore...adopts the following Statement of Affirmations to guide the understanding and response of the agencies, governing bodies, and

people of the church...

The Fundamental Importance of Health. Good health – physical, mental, and spiritual – is both a God-given gift and a social good of special moral importance; one that derives its importance from our biblical and theological heritage and from its effect on the opportunities available to members of society. Good health is a basic need and an essential purpose of human and societal development.

Health Has Many Determinants. Health is determined by what we are born with, how we are nurtured, living conditions, income, education, how we lead our lives, the natural and social environment, access to medical care, our spiritual and psychological state, and our relationships in the communities where we live.

Personal Responsibility for Health. Each person has a moral obligation – a private and public duty – to value and care for his or her own health and the health of the community. We are stewards of God's creation. For most of us, there is ample room to adopt more healthful lifestyles.

Societal Responsibility for Health. Society and its constituent public, private, and voluntary organizations have a duty – a moral obligation – to promote a healthful environment and to assure the availability of health-giving resources to all people. Free markets alone cannot provide for the adequate supply and equitable distribution of these resources. Society's institutions must always strive for the best achievable standards and the most effective performance of the health care system.

Preventive Care. The indispensable foundation on which both individual and societal responsibility for health rests is a consistent major focus on health promotion and maintenance and on preventive care services such as prenatal care, disease control, early detection and diagnosis, mental health services, sex education, and suicide and substance-abuse counseling.

Safeguards Against Unhealthy Working and Living Environments. A community's healthfulness is seriously influenced by the quality of the natural environment and by the interdependent flows of food and materials, energy, and waste products between human beings and nature. Healthy working and living environments are essential to individual and collective health. Clean air, pure water, effective sanitation, nutritious diet, adequate housing, and a safe and nontoxic workplace and living space are all essential to health.

Equal Access to Appropriate and Necessary Care. Every person must have affordable, quality health services. Access should not be limited by income, ethnicity, or geography. It is the proper function of all groups of society, including government, in their concern for justice to ensure equal access to health services .

Responsible Limits. The worship of physical perfection, no less than of worldly wealth, is idolatry. Mortality is an inevitable part of our creation and is the constant backdrop to our efforts to postpone death and overcome disease. A society is justified in placing limits beyond basics on its health care expenditures, balancing them against other needs such as housing, education, employment, and the elimination of poverty. No principle of justice entitles a patient to every conceivable form of beneficial treatment ([4], p. 524).

Social teaching on health care delivery shows the careful line that the Reformed tradition walks between utopianism and nihilism. Moral response to the righteous will and requirement of God, along with the Christ-inspired spirit of sympathy and mercy, prompts us to seek to ameliorate the harshness, misery, and sickness of life by personal ministry and by collective structures of provident care. We do this though we know that perfect healing is not possible in this life, this world. Socialism and socialized medicine are rooted in utopian imagination, not unlike the religious hope for "the Kingdom of God on Earth". Opposed to this is a kind of Darwinism which sees every individual desperately involved in a struggle for survival and self-aggrandizement. At their worst, capitalistic health philosophies stress this philosophy of "get what you can while you can". A disregard verging on nihilism, even therapeutic nihilism, disowns any accountability for the welfare of others.

A kind of wisdom of the middle way is found in Calvinism. Though it knows that sin, sickness, and death will forever be our earthly plight, it knows also that our duty calls us to stand by the widow and the orphan, the sick and the dying, the hungry and the lonely, indeed, anyone in need. Christ is present incognito (*Matthew* 25) in all who need help. Reformed health care ethics is born and built on this conviction.

VI. ISSUES CONCERNING THE END OF LIFE

At the end of 1990, the United States Congress passed a law, which President Bush signed, which mandates that health care institutions funded by federal resources (Medicare, Medicaid) advise patients of their legal rights to make decisions about care in terminal illness. Patients will be told about "advance directives" such as "living wills" and durable powers of attorney. Hospitals, nursing homes, hospices, and other facilities must enter these conversations and decisions into the medical record, and institutions must exhibit procedures to comply with patients' wishes. Such policy has come about because of persistent lobbying by Reformed church bodies, along with many other religious and secular agencies. The resonance of public policy with the following Church policy is evident:

Many members of the Presbyterian Church (U.S.A.) will face health care decisions toward the end of life that they could not have anticipated, and many of those decisions will require judgments that relate to values held by the patient. Therefore, the 195th General Assembly (1983) calls upon its members to:

a. Select their physicians with regard not only to the skillfulness of the medical care that they can provide but also for their values regarding human life and community, whenever such a choice is available.
b. Take time to reflect on their own values and discuss these with family members, close friends, and their clergy.
c. Speak with their physicians about their concerns regarding care and become educated about their conditions in order to permit informed decision-making.
d. Provide instructions (and designate two agents to carry out instructions) with regard to extraordinary therapies and treatments to prolong life.

The church should be a place where individuals and families can make plans about death, manner of death, living wills....[Hence it should:]

a. Request the Program Agency to make available information and study tools for use by congregations regarding options available at the end of life and means of informing health care professionals of these wishes. Advocate that human need and benevolence replace the opportunism and exploitation that so often surround the death and experience presently....Harmony and integration should be sought between intensive care, curative hospitals and hospices so that end of life care can be free from jurisdictional conflict and that therapeutic and palliative care are available to all ([1], p. 367).

While dealing with the general issues of preparation for one's dying and advance directives, the Reformed tradition has not dealt in recent years with the questions of organ replacement, definition of death, and withholding treatment or active euthanasia. Again, the moral theologians of the tradition are active in exploring these questions. Bouma *et al.* devote many pages of their report to the themes of "death and covenantal caring", "dying contrasted with being sick", "competence", "the living will", "allowing to die (passive euthanasia), and killing", "withdrawing nourishment", and "suicide and mercy killing." In keeping with the conservative approach of the Christian Reformed community, the report is cautious about withdrawal of life-support, especially nourishment, and opposed to suicide, assisted suicide, and active euthanasia ([3], p. 268 ff.).

The Reformed medico-moral tradition is the fruit of an ancient "holiness" heritage combined with a modern movement of secular reason. The holiness movement, with its sense of the sanctity of human life and the purity of the medical profession, is found in the Hippocratic and Hebraic practices of healing. This movement and mood in medical practice discountenances abortion and euthanasia, both as violations of the sacred nature and destiny of a human being and as corruptions of the Hippocratic pledge "in holiness [to] practice my art".

The rational heritage argues for more human autonomy and authority over life and death. Rational suicide, especially in the face of debilitating and demeaning disease, has long been a moral conviction of philosophers. Even in Plato and Aristotle (philosophers touched by the holiness tradition), and certainly in the Stoics, Enlightenment philosophers, and moderns like Albert Camus, suicide is seen as tragic necessity, even a courageous act. The Reformed tradition has been shaped by Puritan and Scottish divines, as well as by their skeptical sons, John Locke and David Hume.

A synthesis of these great thanatologies seems to be emerging in the late twentieth century. Remaining at one's post until relieved, laboring in life's vineyard until called home, trusting God, the giver and receiver of our existence, that he will bestow our death in his good time, is probably the cardinal doctrine which bears on these issues. Yet we should not yield to any heteronomous death or anti-death. We should not be willing to relinquish freedom and conscience to necessities such as technological imperatives ("if it's there, it must be used") or so called "medical advances", which inflict us with new deaths as they relieve us of others. Even the blunt necessity of law or economics should not preclude us from dying our own death in due season and giving our life back to our Creator.

Like other religious communions, the Reformed faith tradition is beginning now to give concentrated study to a compelling range of bioethical issues. Task forces on human sexuality and ordination are dealing with agonizing issues such as homosexuality. Other task forces are revisiting abortion and other bioethical issues. Future guidance and policy promises to build upon the foundations here reviewed.

University of Illinois Medical Center
Chicago, Illinois, U.S.A.

BIBLIOGRAPHY

1. Advisory Council on Church and Society: 1983, "The Covenant of Life and Creation" (Report), Minutes of The General Assembly of the United Presbyterian Church (U.S.A.), Louisville, Kentucky.
2. "Blue Book I": 1978, The 19th General Assembly of the United Presbyterian Church in the U.S.A., San Diego, California.
3. Bouma, H. *et al.*: 1989, *Christian Faith, Health, and Medical Practice*, Eerdmans, Grand Rapids.
4. "Life Abundant: Values, Choices and Health Care. The Responsibility and Role of the Presbyterian Church (U.S.A.)": 1989, Minutes of The General Assembly of the United Presbyterian Church (U.S.A.), Louisville, Kentucky.
5. Smylie, J.H.: 1988, "The Reformed Tradition, Health and Healing", in W. Wiest (ed.), *Health Care and Its Costs*, New York, University Press of America.
6. Vaux, K.: 1988, "Abortion in America: A Season for Creative Moral Awareness" (unpublished paper).
7. Vaux, K.: 1984, *Health and Medicine in the Reformed Tradition*, Crossroads, New York.

NOTES ON CONTRIBUTORS

Joseph Boyle, Ph.D., is Professor of Philosophy, St. Michael's College, University of Toronto, Toronto, Ontario, Canada.

Courtney S. Campbell, Ph.D., is Assistant Professor of Religious Studies, Oregon State University, Corvallis, Oregon, U.S.A.

Prakash N. Desai, M.D., is Professor of Clinical Psychiatry, Department of Psychiatry, University of Illinois Medical Center, Chicago, and Chief, Psychiatry Service, Veterans Administration West Side Medical Center, Chicago, Illinois, U.S.A.

Masao Fujii, Ph.D., Professor of Anthropology and Sociology of Religion, Taisho University, Tokyo, Japan.

Judith A. Granbois, M.A., is Program Associate, Poynter Center for the Study of Ethics and American Institutions, Indiana University, Bloomington, Indiana, U.S.A.

Stanley S. Harakas, Th.D., is Professor of Christian Ethics at Holy Cross Greek Orthodox School of Theology in Brookline, Massachusetts, U.S.A.

Hassan H. Hathout, M.D., Ph.D., is Advisor, Islamic Center of Southern California, Los Angeles, California, U.S.A.

Paul T. Nelson, Ph.D., is Assistant Professor, Department of Religion, Wittenberg University, Springfield, Ohio, U.S.A.

Robert L. Shelton, Ph.D., is Associate Professor and Chairman, Department of Religious Studies, University of Kansas, Lawrence, Kansas, U.S.A.

Paul D. Simmons, Ph.D., is Professor of Christian Ethics and Director, Clarence Jordan Center for Christian Ethical Concerns, Southern Baptist Theological Seminary, Louisville, Kentucky, U.S.A.

David H. Smith, Ph.D., is Professor of Religious Studies and Director, Poynter Center for the Study of Ethics and American Institutions, Indiana University, Bloomington, Indiana, U.S.A.

Avraham Steinberg, M.D., is Director, Center for Medical Ethics, Hadassah Medical School, Hebrew University, and Pediatric Neurologist, Department of Pediatrics, Shaare Zedek Medical Center, Jerusalem, Israel.

Kenneth L. Vaux, Th.D., is Professor of Ethics, Department of Medical Education, and Director, Program in Ethics, Department of Internal Medicine, University of Illinois Medical Center, Chicago, Illinois, U.S.A.

GENERAL INDEX

INDEX OF AUTHORS AND DOCUMENTS DISCUSSED

KLUWER ACADEMIC PUBLISHERS
101 Philip Drive • Norwell, MA 02061 • (617) 871-6600
Telex 200190 • FAX (617) 871-6528
E-mail: Kluwer@world.std.com.

October 1, 1991

To the Book Review Editor:

We are pleased to send to you a copy of our newest publication:

TITLE: Bioethics Yearbook. Volume 1: Theological Developments in Bioethics.

AUTHOR (S):

EDITOR (S): Baruch A. Brody, et al.

SERIES:

PRICE: $ 99.00 cloth

ISBN: 0-7923-1280-2

PUBLICATION DATE:

We would appreciate receiving two copies of your review in due course.

Thank you.

Sincerely,

Ann Rollins

Ann Rollins
Marketing Assistant

Incorporated in Massachusetts as Kluwer Boston, Inc.
Kluwer Academic Publishers - The Netherlands   Kluwer Academic Publishers - Boston
Martinus Nijhoff Publishers   Graham & Trotman